Battery Management Systems
Accurate State-of-Charge Indication
for Battery-Powered Applications

Philips Research

VOLUME 9

Editor-in-Chief
Dr. Frank Toolenaar
Philips Research Laboratories, Eindhoven, The Netherlands

SCOPE TO THE *'PHILIPS RESEARCH BOOK SERIES'*

As one of the largest private sector research establishments in the world, Philips Research is shaping the future with technology inventions that meet peoples' needs and desires in the digital age. While the ultimate user benefits of these inventions end up on the high-streeet shelves, the often pioneering scientific and technological basis usually remains less visisble.

This 'Philips Research Book Series' has been set up as a way for Philips researchers to contribute to the scientific community by publishing their comprehensive results and theories in book form.

Dr. Rick Harwig

Battery Management Systems
Accurate State-of-Charge Indication
for Battery-Powered Applications

by

Valer Pop
Holst Centre/IMEC-NL
Eindhoven, The Netherlands

Henk Jan Bergveld
NXP Semiconductors
Eindhoven, The Netherlands

Dmitry Danilov
Eurandom, Eindhoven
The Netherlands

Paul P.L. Regtien
University of Twente
Enschede, The Netherlands

and

Peter H.L. Notten
Philips Research Laboratories Eindhoven
Eindhoven University of Technology
Eindhoven, The Netherlands

Valer Pop
Holst Centre/IMEC-NL
Eindhoven, The Netherlands

Henk Jan Bergveld
NXP Semiconductors
Eindhoven, The Netherlands

Dmitry Danilov
Eurandom
Eindhoven, The Netherlands

Paul P.L. Regtien
University of Twente
Enschede, The Netherlands

Peter H.L. Notten
Philips Research Laboratories Eindhoven
Eindhoven University of Technology
Eindhoven, The Netherlands

ISBN 978-1-4020-6944-4   e-ISBN 978-1-4020-6945-1

Library of Congress Control Number: 2008921935

© 2008 Springer Science+Business Media B.V.
No part of this work may be reproduced, stored in a retrieval system, or transmitted in any form or by any means, electronic, mechanical, photocopying, microfilming, recording or otherwise, without written permission from the Publisher, with the exception of any material supplied specifically for the purpose of being entered and executed on a computer system, for exclusive use by the purchaser of the work.

Printed on acid-free paper.

9 8 7 6 5 4 3 2 1

springer.com

To Raluca, Peggy, Olga, Elly and Pascalle

Table of contents

List of abbreviations	xi
List of symbols	xiii

1. Introduction   1
   1.1 Battery Management Systems   1
   1.2 State-of-Charge definition   3
   1.3 Goal and motivation of the research described in this book   4
   1.4 Scope of this book   6
   1.5 References   7

2. State-of-the-Art of battery State-of-Charge determination   11
   2.1 Introduction   11
   2.2 Battery technology and applications   11
      2.2.1 General operational mechanism of batteries   13
      2.2.2 Battery types and characteristics   14
      2.2.3 Summary   16
   2.3 History of State-of-Charge indication   16
   2.4 A general State-of-Charge system   23
   2.5 Possible State-of-Charge indication methods   24
      2.5.1 Direct measurement   26
      2.5.2 Book–keeping systems   32
      2.5.3 Adaptive systems   34
      2.5.4 Summary   37
   2.6 Commercial State-of-Charge indication systems   38
   2.7 Conclusions   41
   2.8 References   42

3. A State-of-Charge indication algorithm   47
   3.1 An introduction to the algorithm   47
   3.2 Battery measurements and modelling for the State-of-Charge indication algorithm   47
      3.2.1 EMF measurement and modelling   48
      3.2.2 Overpotential measurement and modelling   50
   3.3 States of the State-of-Charge algorithm   52
   3.4 Main issues of the algorithm   54
      3.4.1 EMF measurement, modelling and implementation   55
      3.4.2 Overpotential measurement, modelling and implementation   57
      3.4.3 Adaptive systems   58
   3.5 General remarks on the accuracy of SoC indication systems   59
   3.6 Conclusions   59
   3.7 References   60

4. Methods for measuring and modelling a battery's Electro-Motive Force   63
   4.1 EMF measurement   63
   4.2 Voltage prediction   69

4.2.1 Equilibrium detection	69
4.2.2 Existing voltage-relaxation models used for voltage prediction	70
4.2.3 A new voltage-relaxation model	73
4.2.4 Implementation aspects of the voltage-relaxation model	75
4.2.5 Comparison of results obtained with the different voltage-relaxation models	81
4.2.6 Summary	82
4.3 Hysteresis	**83**
4.4 Electro-Motive Force modelling	**86**
4.5 Conclusions	**93**
4.6 References	**93**
5. Methods for measuring and modelling a battery's overpotential	**95**
5.1 Overpotential measurements	**95**
5.1.1 Overpotential measurements involving partial charge/discharge steps	95
5.1.2 Overpotential measurements involving full (dis)charge steps	100
5.2 Overpotential modelling and simulation	**103**
5.2.1 Overpotential modelling	103
5.2.2 Simulation results	104
5.3 Conclusions	**108**
5.4 References	**109**
6. Battery aging process	**111**
6.1 General aspects of battery aging	**111**
6.1.1 Li-ion battery aging	111
6.1.2 Q_{max} measurements	113
6.2 EMF measurements as a function of battery aging	**114**
6.2.1 The voltage-relaxation model as a function of battery aging	114
6.2.2 EMF GITT measurement results obtained for aged batteries	120
6.2.3 The charge/discharge Electro-Motive Force difference as a function of battery aging	125
6.2.4 EMF modelling as a function of battery aging	130
6.3 Overpotential dependence on battery aging	**132**
6.3.1 Overpotential measurements as a function of aging	132
6.4 Adaptive systems	**137**
6.4.1 Electro-Motive Force adaptive system	137
6.4.2 Overpotential adaptive system	140
6.5 Conclusions	**141**
6.6 References	**142**
7. Measurement results obtained with new SoC algorithms using fresh batteries	**145**
7.1 Introduction	**145**
7.2 Implementation aspects of the algorithm	**146**

7.2.1 A new SoC algorithm	146
7.2.2 Implementation aspects of the SoC algorithm	150
7.3 Results obtained with the algorithm using fresh batteries	**151**
7.4 Uncertainty analysis	**155**
7.4.1 Uncertainty in the real-time SoC evaluation system	155
7.4.2 The SoC uncertainty	158
7.4.3 The remaining run-time uncertainty	161
7.5 Improvements in the new SoC algorithm	**164**
7.5.1 A new State-of-Charge-Electro-Motive Force relationship	164
7.5.2 A new State-of-Charge-left model	165
7.5.3 Determination of the parameters of the new models	166
7.5.4 Test results	169
7.5.5 Uncertainty analysis	173
7.6 Comparison with Texas Instruments' bq26500 SoC indication IC	**174**
7.6.1 The bq26500 SoC indicator	174
7.6.2 Comparison of the two SoC indicators	176
7.7 Conclusions	**178**
7.8 References	**179**
8. Universal State-of-Charge indication for battery-powered applications	**181**
8.1 Introduction	**181**
8.2 Implementation aspects of the overpotential adaptive system	**182**
8.3 SoC=f(EMF) and SoC_l adaptive system	**183**
8.4 Results obtained with the adaptive SoC system using aged batteries	**185**
8.5 Uncertainty analysis	**188**
8.6 Results obtained with other Li-based battery	**189**
8.6.1 EMF and SoC_l modelling results obtained for the Li-based battery	190
8.6.2 Experimental results	196
8.7 Practical implementation aspects of the SoC algorithm	**200**
8.7.1 Hardware design of the evaluation board	200
8.7.2 Software design of the evaluation board	204
8.7.3 Measurement results	205
8.7.4 Boostcharging	208
8.8 Conclusions	**218**
8.9 References	**219**
9. General conclusions	**221**

List of abbreviations

ADC	Analog-to-Digital Converter
ANN	Artificial Neural Networks
AR_d	Actual Rate discharge current
ACU	Adaptive Control Unit
b	Boostcharge
BMS	Battery Management System
BKM	Book-Keeping Module
cc	Coulomb counting
CC	Constant-Current
CCCV	Constant-Current-Constant-Voltage
CCCVCCCV	$(CCCV)^2$
CD	Compact Disc
C_n	Cycle number
ch	Charge
CCA	Charging Current Accumulator
CAC	Compensated Available Charge
d	Discharge
DAC	Digital-to-Analog Converter
DCA	Discharging Current Accumulator
DAQ	Data Acquisition interface card
Dig. I/O	Digital Input/Output pins
EIS	Electrochemical Impedance Spectroscopy
EMF_{dt}	EMF detection method
EMF_f	EMF fitting method
EKF	Extended Kalman Filters
EEPROM	Electrically Erasable Read-Only Memory
EMF_m	Measured EMF data points
f_{sd}	Self-discharge rate estimation rate
GUI	Graphical User Interface
GITT	Galvanostatic Intermittent Titration Technique
HDQ	High-speed single-wire interface
HEV	Hybrid Electrical Vehicles
ISP	In-System Programming
IF	In-Functional
IC	Integrated Circuit
I^2C	Inter-Integrated-Circuit Bus
ICA	Integrated Current Accumulator
JTAG	Joint Test Action Group
KF	Kalman Filters
LED	Light-Emitting Diode
LCD	Liquid-Crystal Display
Li-ion	Lithium-ion
LA	Lead-Acid
Li-ion POL	Lithium-ion Polymer
$Li-SO_2$	Lithium–Sulphur Dioxide
Li	Lithium
LCR	Learning Count Register

LMD	Last Measured Discharge
NiCd	Nickel–Cadmium
NiMH	Nickel–Metal Hydride
NI	National Instruments
NAC	Nominal Available Charge
NTC	Negative-Temperature-Coefficient
N	Number of ADC bits
OLS	Ordinary Least Squares
PGA	Programmable Gain Amplifier
PM	Power Module
ph	Phase transition
PC	Personal Computer
RTOS	Real-Time Operating System
ROM	Read-Only Memory
RAM	Random-Access Memory
SoH	State-of-Health
SBS	Smart Battery System
sd	Self-discharge
SEI	Solid Electrolyte Interface
SCB-68	National Instruments connector board
SPI	Serial Peripheral Interface
UART	Universal Asynchronous Receiver Transmitter
V_{pm}	Voltage prediction model
VDQ	Valid Discharge flag
η_m	Overpotential model

List of symbols

Symbol	Meaning	Value	Unit
a	aged		–
A	Parameter for the SoC-EMF relationship modelling		[1]
a_{10}	Parameter for the SoC-EMF relationship modelling		[1]
a_{11}	Parameter for the SoC-EMF relationship modelling		[1]
a_{12}	Parameter for the SoC-EMF relationship modelling		[1]
a_{20}	Parameter for the SoC-EMF relationship modelling		[1]
a_{21}	Parameter for the SoC-EMF relationship modelling		[1]
a_{22}	Parameter for the SoC-EMF relationship modelling		[1]
c_o	Parameter for overpotential modelling		[Ω A^{-1}]
c_1	Parameter for overpotential modelling		[V]
c_2	Parameter for overpotential modelling	1	[$\sqrt{s}$]
c_3	Parameter for overpotential modelling		[J]
c_4	Parameter for overpotential modelling	1	[A^{-1}]
c_5	Parameter for overpotential modelling		[A]
C_d	Mean discharge C-rate current		–
DoD	Depth-of-Discharge		[%]
DoC	Depth-of-Charge		[%]
EMF	Electro-Motive Force		[V]
E_q^o	Amount of the energy that cannot be obtained from the battery		[J]
E_q^I	Non-linear part of the amount of the energy that cannot be obtained from the battery		[J]
E_{eq}^+	Equilibrium potential of the positive electrode		[V]
E_{eq}^-	Equilibrium potential of the negative electrode		[V]

Symbol	Meaning	Value	Unit
E_0^+	Standard redox potential of the positive electrode		[V]
E_0^-	Standard redox potential of the negative electrode		[V]
EDV_1	End-of Discharge Voltage level		[V]
EMF_m^{es}	Estimated EMF model		[V]
$EMF_{a5.4\%}$	EMF obtained for the 5.4% capacity loss battery (a=aged)		
$EMF_{a25.4\%}$	EMF obtained for the 25.4% capacity loss battery (a=aged)		
EMF_1	First EMF point determined for the EMF adaptation		[V]
EMF_{ad}	EMF retrieved by means of the adaptation method		[V]
E_0	Parameter for the SoC-EMF relationship modelling		[V]
EMF_p	Predicted EMF voltage		[V]
EMF_{GITT}	EMF measurement by means of the GITT		
EMF_f	Fitted EMF		[V]
F	Faraday constant	96485	[C mol^{-1}]
f	Frequency		[Hz]
f	Fresh		–
I_s^{max}	Constant maximum current (s=standard)		[A]
I_s^{min}	Predefined minimum current (s=standard)		[A]
I_b^{max}	Maximum boostcharging current		[A]
I	Current that flows into(out) of the battery		[A]
I_M	Uncertainties from Maccor current measurements	Neglected	[A]
I_s	Standby current	$-1\ 10^{-3}$	[A]
I_b	Backlight-on state current		[A]
I_d	Measured discharge current		[A]
I_{lim}	Limit current		[A]
M	Slope of second asymptote of prior-art voltage-prediction method		–
n_0	Parameter related to the magnitude of the diffusion overpotential		[1]
n_1	Parameter related to the magnitude of the diffusion overpotential		[T^{-1}]

List of symbols

Symbol	Meaning	Value	Unit
OCV	Open-Circuit Voltage		[V]
OCV_f	Battery OCV for a fresh battery		[V]
OCV_a	Battery OCV for an aged battery		[V]
$par(T_{ref})$	Value of one of the EMF-SoC model parameters at temperature T_{ref}		[1]
p_{11}	Parameter for the SoC-EMF relationship modelling		[1]
p_{12}	Parameter for the SoC-EMF relationship modelling		[1]
p_{21}	Parameter for the SoC-EMF relationship modelling		[1]
p_{22}	Parameter for the SoC-EMF relationship modelling		[1]
q_{11}	Parameter for the SoC-EMF relationship modelling		[1]
q_{12}	Parameter for the SoC-EMF relationship modelling		[1]
q_{21}	Parameter for the SoC-EMF relationship modelling		[1]
q_{22}	Parameter for the SoC-EMF relationship modelling		[1]
Q_{in}	Charge present in the battery at the time t		[Ah]
Q_{d2}	Discharge capacity for battery number 2		[Ah]
Q_{d1}	Discharge capacity for battery number 1		[Ah]
Q_{d1}^{1}	Q_{d1} value after the first cycle	$1165\ 10^{-3}$	[Ah]
Q_{d1}^{220}	Q_{d1} value after 220 cycles	$675\ 10^{-3}$	[Ah]
Q_{d2}^{1}	Q_{d2} value after the first cycle	$1150\ 10^{-3}$	[Ah]
Q_{d2}^{2000}	Q_{d2} value after 2000 cycles	$935\ 10^{-3}$	[Ah]
Q_{max}	Battery maximum capacity		[Ah]
Q_{max1}	Maximum capacity of battery 1	$875\ 10^{-3}$	[Ah]
Q_{max2}	Maximum capacity of battery 2	$1110\ 10^{-3}$	[Ah]
Q_{loss}	Capacity loss		[%]
Q_d	Discharge battery capacity		[Ah]
Q_{dd}	Decrease in Q_d		[%]
Q_{max}^{+}	Maximum capacity of the positive electrode	1.00	[1]
Q_{max}^{-}	Maximum capacity of the negative electrode	$4.23\ 10^{-1}$	[1]

List of symbols

Symbol	Meaning	Value	Unit
Q_{max}	Amount of the electrochemically active Li^+ ions inside the battery	8.11 10^{-1}	[1]
Q_{maxr}	Reference maximum capacity value	1173 10^{-3}	[Ah]
Q_0^-	Amount of the Li^+ ions that will remain in the negative electrode after discharging	9.63 10^{-4}	[1]
Q_{ch}	Amount of charge flowing into the battery during the charge state		[Ah]
Q_d	Discharge capacity		[Ah]
R	Gas constant	8.314	J(mol K)$^{-1}$)
R	Resistance		[Ω]
R_d^0	Linear contribution of the "diffusion" resistance		[Ω]
R_d^I	Non-linear contribution of the "diffusion" resistance		[Ω]
R_S	Sense resistor	20 10^{-3}	[Ω]
$R_{\Omega k}$	Linear contribution of the "ohmic" and "kinetic" resistance		[Ω]
R_{Ik}	Non-linear contribution of the "ohmic" and "kinetic" resistance		[Ω]
R_{HF}	High-frequency resistance		[Ω]
s_x	Sign of parameter x		–
SoC	State-of-Charge		[%]
SoC(EMF)	SoC calculated based on the EMF voltage		[%]
SoC(V_p)	SoC calculated based on the predicted voltage		[%]
SoC_l	SoC-left		[%]
SoC_{lm}^{es}	Estimated SoC_l model		[%]
SoC_{lm}	Measured SoC_l		[%]
SoC_{lf}	Fitted SoC_l		[%]
SoC_e	SoC error		[%]
SoC_{si}	Initial SoC in the standby state		[%]
SoC_{sf}	Final SoC in the standby state		[%]
SoC_{st}	SoC indicated at the start		[%]
SoC_{end}	SoC indicated at the end		[%]
SoC_{in}	SoC in the initial state		[%]
SoC_s	SoC in the standby state		[%]
SoC_t	SoC in the transitional state		[%]
SoC_{ch}	SoC in the charge state		[%]
SoC_d	SoC in the discharge state		[%]
SoC_b	SoC in the backlight-on state		[%]

List of symbols   xvii

Symbol	Meaning	Value	Unit
$SoC_e(V_p)$	Error of the SoC calculated based on a predicted voltage		[%]
SoC_e (OCV)	Error of the SoC calculated based on the OCV		[%]
t	time		[s]
T	Temperature		[°C]
T_{ref}	Reference temperature	25	[°C]
t_b	Time boostcharging		[min]
t_{empty}	Time where battery is empty		[min]
t_r	Remaining run-time		[min]
t_{rst}	Remaining run-time at the start of experiment		[min]
t_{re}	Error in the remaining run-time at the end of the experiment		[min]
t_{rstp}	Predicted remaining run-time at the start of the experiment		[min]
t_{rstm}	Measured remaining run-time at the start of the experiment		[min]
t_{rre}	Relative error in the remaining run-time		[%]
T_M	Uncertainties from Maccor temperature measurements	Neglected	[°C]
t_M	Uncertainties from Maccor time measurements	Neglected	[s]
U_1^+	Interaction energy coefficient of the positive electrode	$9.68\ 10^2$	[1]
U_2^+	Interaction energy coefficient of the positive electrode	$1.63\ 10^4$	[1]
U_1^-	Interaction energy coefficient of the negative electrode	$-6.92\ 10^3$	[1]
U_2^-	Interaction energy coefficient of the positive electrode	$-6.89\ 10^3$	[1]
U_j^-	Interaction energy coefficient of the negative electrode		–
U_j^+	Interaction energy coefficient of the positive electrode		–
V_{bat}	Battery terminal voltage		[V]
V	Battery voltage		[V]
V_d	Battery voltage after discharging		[V]
V_{ch}	Battery voltage after charging		[V]
V_M	Uncertainties from Maccor voltage measurements	Neglected	[V]
V_o	Battery voltage at t=1min		[V]

Symbol	Meaning	Value	Unit
V_p	Predicted voltage		[V]
V_{EoD}	End-of-Discharge voltage	3	[V]
V_{ocf}	Fully stabilized Open-Circuit Voltage		[V]
V_b^{max}	Maximum boostcharging voltage		[V]
V_s^{max}	Maximum charge voltage in CV mode	4.2	[V]
V_∞	Final relaxation voltage		[V]
V_t	Relaxation voltage at time t		[V]
V_m	Measured voltage in the first 5 minutes of the relaxation		[V]
V_{off}	Voltage offset		[V]
V_{abs}	Absolute voltage accuracy at full scale	$0.95\ 10^{-3}$	[V]
V_{fs}	Full-scale range of the ADC voltage		[V]
x	Parameter for the SoC-EMF relationship modelling		–
X	Exponent of time		–
x_{Li}	Molfraction of Li$^+$ ions inside the positive electrode		[1]
X_p	Value of X at asymptotes intersection	1.64	–
w_2	Parameter for the SoC-EMF relationship modelling		[1]
Z	Battery impedance		[Ω]
z_{Li}	Molfraction of the Li ions inside the negative electrode		–
z_{ph}	Phase transition in the negative electrode	$4.44\ 10^{-1}$	[1]
x_{ph}	Phase transition in the positive electrode	$8.44\ 10^{-1}$	[1]
ΔU_1^+	Sensitivity to temperature of parameter U_1^+	–1.89	[T^{-1}]
ΔU_2^+	Sensitivity to temperature of parameter U_2^+	$5.14\ 10^1$	[T^{-1}]
ΔU_1^-	Sensitivity to temperature of parameter U_1^-	$2.18\ 10^1$	[T^{-1}]
ΔU_2^-	Sensitivity to temperature of parameter U_2^-	$2.16\ 10^1$	[T^{-1}]

List of symbols

Symbol	Meaning	Value	Unit
$\Delta\zeta_1^+$	Sensitivity to temperature of parameter ζ_1^+	$2.22\ 10^{-4}$	$[T^{-1}]$
$\Delta\zeta_1^-$	Sensitivity to temperature of ζ_1^-	$2.04\ 10^{-4}$	$[T^{-1}]$
Δx_{ph}	Sensitivity to temperature of x_{ph}	$1.58\ 10^{-6}$	$[T^{-1}]$
Δz_{ph}	Sensitivity to temperature of z_{ph}	$1.37\ 10^{-4}$	$[T^{-1}]$
$\Delta\sigma_{x1}$	Sensitivity to temperature of σ_{x1}	$-2.24\ 10^{-6}$	$[T^{-1}]$
$\Delta\sigma_{z1}$	Sensitivity to temperature of σ_{z1}	$8.98\ 10^{-4}$	$[T^{-1}]$
ΔQ_0^-	Sensitivity to temperature of Q_0^-	$4.34\ 10^{-5}$	$[T^{-1}]$
ΔA	Sensitivity to temperature of A		$[T^{-1}]$
Δw_2	Sensitivity to temperature of w_2		$[T^{-1}]$
ΔV_{max}	Maximum error in the voltage measurement		[V]
Δa_{20}	Sensitivity to temperature of a_{20}		$[T^{-1}]$
Δa_{10}	Sensitivity to temperature of a_{10}		$[T^{-1}]$
Δa_{11}	Sensitivity to temperature of a_{11}		$[T^{-1}]$
Δa_{12}	Sensitivity to temperature of a_{12}		$[T^{-1}]$
Δp_{11}	Sensitivity to temperature of p_{11}		$[T^{-1}]$
Δp_{12}	Sensitivity to temperature of p_{12}		$[T^{-1}]$
Δp_{21}	Sensitivity to temperature of p_{21}		$[T^{-1}]$
ΔE_0	Sensitivity to temperature of E_0		$[V\ T^{-1}]$
Δpar	Sensitivity to temperature determined for each parameter par(T_{ref})		$[T^{-1}]$
α	Rate-determining variable		–
γ	Rate-determining variable		–
δ	Rate-determining variable		–
δ	Parameter in the voltage prediction model		–
ε_t	Random error term		–
Γ	Constant for voltage prediction	+1 (discharge) −1 (charge)	–
ζ_j^+	Constant in the positive electrode		–
ζ_j^-	Constant in the negative electrode		–
ζ_1^+	Constant in the positive electrode	0.00	[1]

Symbol	Meaning	Value	Unit
ζ_1^-	Constant in the negative electrode	$-2.14 \; 10^{-5}$	[1]
η_{ch}	Charge overpotential		[V]
η_d	Discharge overpotential		[V]
η_{meas}	Measured overpotential from V and EMF		[V]
η_{calc}	Calculated overpotential		[V]
η_f	Overpotential model parameters for a fresh battery		
η_{df}	Measured discharge overpotential for fresh battery		[V]
η_{chf}	Measured charge overpotential for fresh battery		[V]
η_{ch}^a	Measured charge overpotential for an aged battery		[V]
η_{ch}^f	Measured charge overpotential for a fresh battery		[V]
η_d^a	Measured discharge overpotential for an aged battery		[V]
η_d^f	Measured discharge overpotential for a fresh battery		[V]
ϕ	Phase angle		[rad]
Φ	Standard normal cumulative distribution function		–
τ_q	Time constant associated with the increase in overpotential in an almost empty battery		[s]
τ_d	"Diffusion" time constant		[s]
σ_{x1}	Parameter that determine the smoothness of the phase transition for the positive electrode	$1.72 \; 10^{-2}$	[1]
σ_{z1}	Parameter that determine the smoothness of the phase transition for the negative electrode EMF	$1.22 \; 10^{-1}$	[1]
τ	Voltage relaxation time		[s]

Chapter 1
Introduction

This chapter gives general information on Battery Management Systems (BMSs) and State-of-Charge (SoC) indication that will be required as a background in later chapters. A general block diagram of a BMS is shown in section 1.1. One of the main tasks of a BMS is to keep track of a battery's SoC, which is the main subject of this book. Section 1.2 gives a definition of SoC indication and discusses its importance. The goal and motivation of the research described in this book are discussed in section 1.3. Finally, section 1.4 presents the contents of this book.

1.1 Battery Management Systems

Battery-powered electronic devices have become ubiquitous in modern society. The recent rapid expansion of the use of portable devices (*e.g.* portable computers, personal data assistants, cellular phones, shavers, *etc.* (see Fig. 1.1)) and Hybrid Electrical Vehicles (HEVs) creates a strong demand for fast deployment of battery technologies at an unprecedented rate [1].

Fig. 1.1. Examples of portable devices on the market [2].

The design of a battery-powered device requires many battery-management features, including charge control, battery-capacity monitoring, remaining run-time information, charge-cycle counting, *etc.* For it to be able to offer high precision, each part of the system must be near to perfection. The basic task of a BMS can be defined as follows [1]:

The basic task of a Battery Management System (BMS) is to ensure that optimum use is made of the energy inside the battery powering the portable product and that the risk of damage to the battery is prevented. This is achieved by monitoring and controlling the battery's charging and discharging process.

A general block diagram of a BMS is shown in Fig. 1.2. The basic task of the power module (PM) is to charge the battery by converting electrical energy from the mains into electrical energy suitable for charging the battery. The PM can either be a separate device, such as a travel charger, or it can be integrated within the portable device, as, for example, in shavers [1]. A protection Integrated Circuit (IC) connected in series with the battery is generally needed for lithium-ion (Li-ion) batteries. The reason for this is that battery suppliers are particularly concerned about safety issues due to liability risks. The battery voltage, current and temperature have to be monitored and the protection IC ensures that the battery is never operated under unsafe conditions. The battery manufacturer determines the operating conditions under which it is assumed to be safe to use Li-ion batteries. Outside the safe region, destructive processes may take place [1].

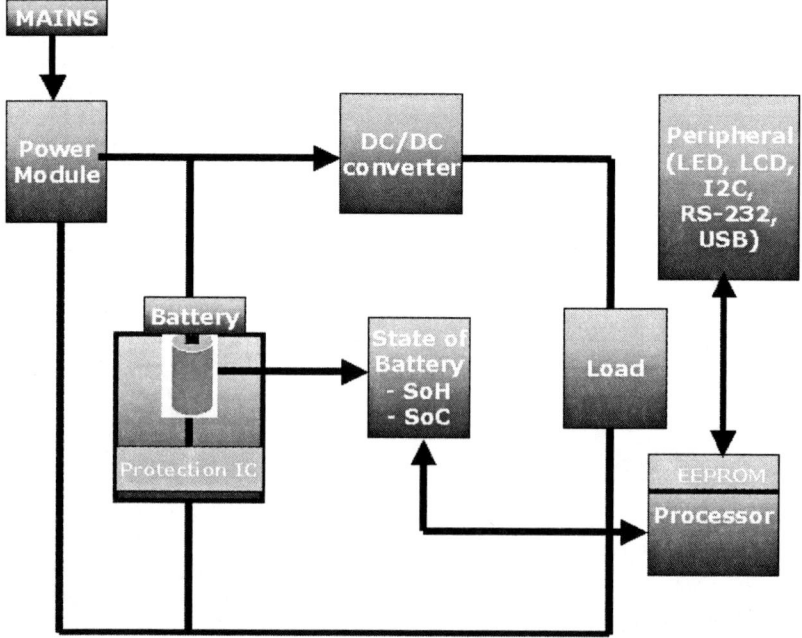

Fig. 1.2. General architecture of a Battery Management System.

The DC/DC converter is used to efficiently condition the unregulated battery voltage (3–4.2 V in Li-ion chemistry) for compatibility with stringent load requirements (see Fig. 1.2). The basic task of the load is to convert the electrical energy supplied by a battery into an energy form that will fulfil the load's function, such as mechanical energy, light, sound, heat, EM radiation, *etc*. The battery status can be indicated in one light-emitting diode (LED) or several such diodes connected in series or on a liquid-crystal display (LCD) that indicates the SoC and the battery's condition (*e.g.* the State-of-Health (SoH)) [1]. The processor is used to run the battery-management software, including the SoC algorithm (see Fig. 1.2). Communication between the BMS and other devices is another important task of the BMS. Depending on the application, various systems can be used for data exchange, such as an inter-integrated-circuit bus interface (I^2C) or some other form of serial interface (see Fig. 1.2). The battery state is used as an input parameter for the portable device's electrical management and it is an important parameter for the user. The battery state can be used to estimate the battery's expected lifetime. It can be simply described by two parameters: SoC and SoH. Both parameters depend on each other and influence the battery performance. More information on these two parameters will be given in the next section.

1.2 State-of-Charge definition

Three terms are relevant with respect to accurately implementing the monitor function of the battery state in a Battery Management System. These three terms are the State-of-Charge (SoC), the State-of-Health (SoH) and the remaining run-time (t_r). The SoC can be defined as follows:

State-of-Charge (SoC) is the percentage of the maximum possible charge that is present inside a rechargeable battery.

The SoC indication involves battery measurements and modelling [1]. As a simple example the battery voltage (V) can be measured and the *V-SoC* relationship can be stored in a look-up table function in a microcontroller [3]–[6]. The size and accuracy of the look-up tables in SoC indication systems depend on the number of stored values, *i.e.* the number of stored *V-SoC* data points. A problem is that the battery voltage changes with temperature, discharge rates and aging. Making the look-up table temperature and discharge-rate dependent can solve the first two dependencies [7]. However, aging of the battery is a complex process that involves many battery parameters (*e.g.* impedance, capacity). The process is too complex to be tackled with simple look-up table implementation [7].

The State-of-Health (SoH) is an indication of the point that has been reached in the battery's life cycle and is a measure of its condition relative to a fresh battery. The SoH can be defined as follows:

State-of-Health (SoH) is a 'measure' that reflects the general condition of a battery and its ability to deliver the specified performance in comparison with a fresh battery.

The SoH indication may involve for example cycle-counting. In the simplest case the number of full charge/discharge cycles (C_n) can be counted and the SoH can be calculated on the basis of a stored *maximum capacity*-C_n function [1]. However, a user doesn't always wait until a battery reaches an empty or a full SoC state. The system should therefore also take into consideration SoC levels other than "empty" and "full", *e.g.* levels defined by the last discharge/charge SoC value before a user starts charging/discharging. Another problem is the spread in both battery and user behaviour. Due to this spread the SoH evolution will be different for each user and application, and will consequently be rather unpredictable. It is not possible to deal with such unpredictable behaviour with a simple charge/ discharge cycles counting implementation. An adaptive system must therefore be used to ensure an accurate SoC indication when the battery ages. Examples of possible adaptive systems are neural networks [8], [9], Kalman filters [10]–[13] and fuzzy logic [14].

The SoC is usually displayed to the user in a graphic bar or in [%]. In the latter case 100% implies a full battery state and 0% the empty state. However, for a user it is convenient to know how long a portable device battery will still be able to deliver power. An SoC indication with a couple of bars does not provide sufficient information. The remaining time-of-use indication will be the most interesting and attractive solution for a portable device user. The remaining run-time can be defined as follows [1]:

The remaining run-time (t_r) is the estimated time that the battery can supply current to a portable device under valid discharge conditions before it will stop functioning.

The remaining run-time can be inferred from the remaining capacity in two ways, depending on the type of load: in the case of a current-type load the remaining capacity in mAh, so expressed as charge, is divided by the drawn current in mA and in the case of a power-type load the remaining capacity in mWh, so expressed as energy, is divided by the drawn power in mW [1]. In this book only current-type loads will be considered for simplicity.

To conclude, an accurate SoC and run-time determination method combined with a SoH calculation will improve a battery's performance and reliability, and will ultimately lengthen its lifetime.

1.3 Goal and motivation of the research described in this book

Accurate SoC and remaining run-time indication for portable devices is important for user convenience and for prolonging the lifetime of batteries. In a survey conducted by market research group TNS involving almost 7,000 mobile users in 15 countries, over 75% of respondents said better battery life is the main feature they want from a future converged device [15]. This motivates the request for an accurate and reliable SoC, run-time and SoH indicator system in portable applications. At the moment Li-ion is the most commonly used battery chemistry in portable applications. Therefore, the focus in this book is on SoC indication for Li-ion batteries. The chosen battery in this work is Sony's US18500G3 Li-ion battery.

Accurate SoC information allows a battery to be used within its design limits, so the battery pack does not need to be over-engineered. This allows the use of a smaller, lighter battery, which costs less. However, many examples of poor accuracy and reliability are found in practice. This can be pretty annoying, especially when a portable device suddenly stops functioning whereas sufficient battery capacity is indicated. Poor reliability of the SoC indication system may induce the use of only part of the available battery capacity. For example, a user may be inclined to recharge a battery every day, even when enough battery capacity is indicated on the portable device. This will lead to more frequent recharging than strictly necessary, which in turn leads to earlier wear-out of the battery. The effect of inaccuracy of SoC indication may be even worse when the SoC value is also used to control charging. The battery is either not fully charged or it is overcharged. In the former case, the battery will be recharged more often than needed, which will lead to earlier wear-out. In the latter case, frequent overcharging will lead to a lower cycle life [1].

Many leading semiconductor companies (*e.g.* Philips, Texas Instruments, Microchip, Maxim, *etc.*) are paying more and more attention to accurate State-of-Charge (SoC) indication. Following the technological revolution and the appearance of more power-consuming devices on the automotive electronics and portable devices markets (*e.g.* Third-Generation cellular phones) the simple SoC indication systems of the thirties based on voltage and temperature measurements [16]–[19] have been replaced by more complicated and accurate SoC systems [1], [20]–[27].

Of these, the system presented by Bergveld *et al.* in 2000 [1], [25]–[27], implementing the mathematical models described in [1], has been found to be the most accurate [28], [29]. The developed method refers to SoC estimation for a Li-ion battery. The method is based on current measurement and integration during the charge and discharge state (referred to as Coulomb counting [1], [7], [30]) and Electro-Motive Force (EMF) measurements during the equilibrium state (state in which no current is flowing into or out of the battery and the battery is fully stabilized) [1]. The EMF method is also used for calibrating the SoC system, because the same SoC level in percentage has been found for a certain measured EMF irrespective of the age and temperature of the battery. This calibration is important, because in charge and discharge states the calculated SoC will eventually drift away from the real value due to *e.g.* measurement inaccuracy in the current and the integration over time of this inaccuracy [1]. Apart from simple Coulomb counting, the effect of the overpotential is also considered in the discharge state. Due to this overpotential, the battery voltage during discharge is lower than the EMF. The value of the overpotential depends on the discharge current, SoC, age and temperature.

A couple of shortcomings of the developed method have been revealed [1]. In the first place, the implementation of an accurate battery model is essential for accurate SoC indication. The model applied in the proposed SoC indication system describes the battery EMF and overpotential behaviour, neither of which can be measured directly. Drawbacks of the measurement methods described in [1] have been discussed in [34], [35] along with possible solutions. Secondly, false entries in the equilibrium state have been detected [1]. They influence the EMF estimation and the system's calibration accuracy [1]. Furthermore, the overpotential model presents parameters that are temperature and age dependent. A better model that includes temperature and age dependence needs to be developed.

A designer of a BMS in a portable device will also be interested in the implementation requirements of the mathematical functions used in the SoC indication algorithm in a practical application. Close quantitative agreement between the results of laboratory simulations using the battery models and measurements on a real-time system in which the SoC system is implemented is then of course important. Part of the research described in this book is devoted to optimising such quantitative agreement. The optimised SoC system implementation must agree with the portable device hardware speed and memory requirements. Part of the research described in this book is aimed at finding an optimum implementation method of the SoC algorithm in a real-time system.

The final aim of the method presented in this book is to design and test an SoC indication system capable of predicting the remaining capacity of any Li-based battery and the remaining run-time with an accuracy of 1 minute or better under all realistic user conditions, including aging, a wide variety of load currents and a wide temperature range. The designed SoC indication system must moreover agree with the portable application hardware and software requirements.

1.4 Scope of this book

This book presents the results of research into battery measurements and electrochemical modelling obtained by combining the expertise of electrical engineering with that of electrochemical and computer science. The result is an adaptive method for indicating the SoC and remaining run-time that can be applied to all Lithium systems [31]–[35].

This book is organized as follows. Chapter 2 presents an overview of battery technologies and the state-of-the-art State-of-Charge (SoC) methods. The general operational mechanism of batteries and information on existing SoC indication methods will be discussed. This general information is required as a background in the remaining chapters of this book.

There are several practical methods available for SoC indication. Special attention will be paid to the SoC indication system presented by Bergveld *et al.* in [1], [25], [26], which represents the starting point of this book. This SoC system has been chosen because it is one of the most advanced SoC systems so far proposed in the literature [28], [29]. The main advantages and drawbacks of this system will be presented in Chapter 3. In chapters 4–7 improvements on this algorithm will be presented.

A complete study of a better EMF determination method developed in-house will be presented. This method will lead to a better understanding of the EMF dependence on temperature and aging and of new topics such as the EMF hysteresis. A new EMF model that includes temperature dependence will be developed. This will enable the use of the EMF at different temperatures, which will finally improve the SoC indication accuracy. A new model that predicts the EMF from the first minutes of relaxation will also be presented. Accurate EMF prediction is important because the EMF will also be used as a calibration method in our system. These efforts will be described in chapter 4.

The main drawback of the EMF method is that it does not provide continuous indication of the SoC. Therefore the SoC algorithm also uses Coulomb counting and overpotential prediction. A complete study of a new overpotential determination method also developed in-house will be presented. This method will lead to a better understanding of new topics such as overpotential symmetry and overpotential dependence on C-rate current and aging. A new overpotential model that includes C-rate current dependence will be developed. This will enable the use of the overpotential at different currents and will finally improve the SoC indication accuracy. These efforts will be presented in chapter 5.

Introduction

The main problem of designing an accurate SoC indication system is the unpredictability of both battery and user behaviour. This necessitates the use of an adaptive system. In addition to better measurements and battery models, on-line adaptive solutions will be added to the SoC system. The EMF and overpotential models for example include a variety of parameters that change during cycling of the battery. Innovative methods for adapting the EMF and overpotential model parameters used in the system, including aging effects, will be described in chapter 6.

Chapter 7 will focus on implementation aspects of the algorithm. For testing and evaluation purposes the SoC indication system has been implemented in a real-time laboratory set-up. Simulation results of the mathematical models and experimental results of the implemented SoC system will be presented. The experimental results will show the effectiveness of the presented novel approach for improving the SoC indication accuracy.

The "dream" of the last 70 years of research in the field of SoC is to design an accurate SoC system that will adapt on-line to any type of battery [7]. An innovative algorithm solution that incorporates on-line adaptation solutions will be presented in chapter 8. This algorithm will bring the dream of any researcher in this field nearer to realization. In addition, a demonstration board that incorporates a mobile phone hardware platform has been developed. The applicability and usability of the SoC algorithm for a newly developed ultra-fast recharging algorithm will also be presented in Chapter 8.

Conclusions will be drawn in chapter 9 and recommendations will be made for further research in the interesting field of SoC indication.

1.5 References

[1]   H.J. Bergveld, W.S. Kruijt, P.H.L. Notten, Battery Management Systems – Design by Modelling, Philips Research Book Series, **1**, Kluwer Academic Publishers, Boston (2002)

[2]   V. Pop, H.J. Bergveld, P.H.L. Notten, P.P.L. Regtien, State-of-Charge Indication, Internal Philips Presentation (2005)

[3]   K. Muramatsu, Battery condition monitor and monitoring method, US Patent 4,678,998, filed 9 December (1985)

[4]   E. Peled, H. Yamin, I. Reshef, D. Kelrich, S. Rozen, Method and apparatus for determining the state-of-charge of batteries particularly lithium batteries, US Patent 4,725,784, filed 10 September (1984)

[5]   U. Kopmann, Method of and apparatus for monitoring the state of charge of a rechargeable battery, US Patent 4,677,363, filed 30 June (1987)

[6]   L. Bowen, R. Zarr, S. Denton, A microcontroller-based intelligent battery system, IEEE AES System Magazine, 16–19 (1994)

[7]   V. Pop, H.J. Bergveld, P.H.L. Notten, P.P.L. Regtien, State-of-the-art of State-of-Charge determination, Measurement Science and Technology Journal, **16**, R93–R110 (2005)

[8]   O. Gerard, J.N. Patillon, F. d'Alche-Buc, Neural network adaptive modelling of battery discharge behaviour, Lect. Notes Comput. Sci., **1327**, 1095–1100 (1997)

[9]   S. Grewal, D.A. Grant, A novel technique for modelling the state of charge of lithium ion batteries using artificial neural networks, Proc. Int. Telecommunications, Energy Conf. (IEEE), **484**, 14–18 (2001)

[10] J. Garche, A. Jossen, Battery management systems (BMS) for increasing battery life time, Telecommunications Energy Special, **3**, 81–84 (2000)
[11] G. Plett, Extended Kalman filtering for battery management systems of LiPB-based HEV battery packs: Part 1. Background, J. Power Sources, **134**, 252–261 (2004)
[12] G. Plett, Extended Kalman filtering for battery management systems of LiPB-based HEV battery packs: Part 2. Modeling and identification, J. Power Sources, **134**, 262–276 (2004)
[13] G. Plett, Extended Kalman filtering for battery management systems of LiPB-based HEV battery packs: Part 3. State and parameter estimation, J. Power Sources, **134**, 277–292 (2004)
[14] A.J. Salkind, C. Fennie, P. Singh, T. Atwater, D.E. Reisner, Determination of state-of-charge and state-of-health of batteries by fuzzy logic methodology, J. Power Sources, **80**, 293–300 (1999)
[15] J. Walko, Mobile phone users demand decent batteries, EET Times, (2005)
[16] B.F.W.H. Heyer, One meter battery tester, US Patent 2,225,051, filed May 16 (1938)
[17] H. Dreer, Curtis wheelchair battery fuel gauge, Product Test Report Marketing Services, Dept. Curtis Instruments Inc., (1984)
[18] J.J. Kauzlarich, Electric wheelchair fuel gauge tests, Report No UVA-REC-102-86, UVA Rehabilitation Engineering Center (1986)
[19] R.A. York, Self-testing battery discharge indicator, US Patent 3,932,797, filed 24 December (1974)
[20] Texas Instruments, High-performance battery monitor IC with Coulomb counter, voltage, and temperature measurements, Doc. I.D. SLUS521A (2002)
[21] Texas Instruments, Single-cell Li-ion and Li-Pol battery gas gauge IC for handheld applications (bq junior family), Doc. I.D. SLUS567A (2003)
[22] Texas Instruments, High-performance battery monitor with Coulomb counting and flash memory, Doc. I.D. SLUS509 (2001)
[23] Power Smart, P3 cell model explanations, Application Note 450799 (1999)
[24] Dallas Semiconductors, Characterizing a Li+ cell for use with a fuel cell, Application Note 2412 (2003)
[25] H.J. Bergveld, H. Feil, J.R.G.C.M. Van Beek, Method of predicting the state of charge as well as the use time left of a rechargeable battery, US Patent 6,515,453, filed 30 November (2000)
[26] F.A.C.M. Schoofs, W.S. Kruijt, R.E.F. Einerhand, S.A.C. Hanneman, H.J. Bergveld, Method of and device for determining the charge condition of a battery, US Patent 6,420,851, filed 29 March (2000)
[27] P.H.L Notten, H.J. Bergveld, W.S. Kruijt, Battery Management System and Battery Simulator, Patent US6016047, filled 11 October (1997)
[28] V. Pop, H.J. Bergveld, P.H.L. Notten, P.P.L. Regtien, State-of-Charge Indication in Portable Applications, IEEE International Symposium on Industrial Electronics, **3**, 1007–1012 (2005)
[29] V. Pop, H.J. Bergveld, P.H.L. Notten, P.P.L. Regtien, Smart and Accurate State-of-Charge Indication in Portable Applications, Power Electronics and Drive Systems, **1**, 262–267 (2005)
[30] J.H. Aylor, A. Thieme, B.W. Johnson, A battery state-of-charge indicator for electric wheelchairs, IEEE Trans. Indust. Electron., **39**, 398–409 (1992)

[31] V. Pop, H.J. Bergveld, P.H.L. Notten, P.P.L. Regtien, A Real-Time Evaluation System for a State-of-Charge Indication Algorithm, Proceedings of the Joint International IMEKO TC1+ TC7 Symposium, **1**, 104–107 (2005)
[32] H.J. Bergveld, V. Pop, P.H.L. Notten, Method of estimating the State-of-Charge and of the use time left of a rechargeable battery, and apparatus for executing such a method, Patent WO2005085889 A1 filed February 23 (2005)
[33] H.J. Bergveld, V. Pop and P.H.L. Notten, Apparatus and method for determination of the state-of-charge of a battery when the battery is not in equilibrium, Patent PH006795EP1, filed October 30 (2006)
[34] V. Pop, H.J. Bergveld, J.H.G. Op het Veld, D. Danilov, P.P.L. Regtien, P.H.L. Notten, J. Electrochem. Soc., **153** (11), A2013–A2022 (2006)
[35] Chapter 7 of this book (2007)

Chapter 2
State-of-the-Art of battery State-of-Charge determination

Improvements over time in the use of battery technology and SoC indication will be presented in this chapter. The goal of all the presented SoC determination methods is to arrive at an SoC indication system capable of providing an accurate SoC indication under all realistic user conditions, including those of spread in both battery and user behaviour, a large temperature and current range and aging of the battery. This chapter is organized as follows. Section 2.2 briefly describes the general operational mechanism of batteries and the characteristics of the best-known batteries with their applications. The history of SoC indication is presented in section 2.3. A general State-of-Charge system and possible State-of-Charge indication methods are discussed in sections 2.4 and 2.5, respectively. Section 2.6 focuses on commercially available State-of-Charge systems. Finally, section 2.7 presents concluding remarks.

2.1 Introduction

Humanity has depended on electricity ever since it was first discovered. Without this phenomenon many technological advancements would not have been made. When the need for mobility increased, people switched to portable energy storage devices – first of all wheeled applications, then portable ones and finally wearable use. Several types of rechargeable battery systems are available on the market, such as lead–acid (LA), nickel–cadmium (NiCd), nickel–metal hydride (NiMH), lithium-ion (Li-ion) and lithium-ion polymer (Li-ion POL) batteries. The most important of them will be discussed in this chapter. For almost as long as rechargeable batteries have existed, systems that are capable of indicating the SoC have been around. Several methods for determining the SoC of a battery are known in the art, such as *direct measurements, book-keeping* and *adaptive systems* [1]. An accurate SoC determination method and an understandable and reliable SoC user display will improve a battery's performance and reliability and will ultimately lengthen its lifetime.

2.2 Battery technology and applications

As awkward and unreliable as the early batteries may have been, our descendants may one day look at today's technology in very much the same way as we now view our predecessors' clumsy experiments of 100 years ago [2]. This section focuses on developments in battery technology and characteristics.

In 1800 Volta discovered that a continuous flow of electrical force was generated when certain fluids were used as ionic conductors to promote an electrochemical reaction between two metals or electrodes [2]. This led to the invention of the first voltaic cell, better known as a battery.

In 1859 the French physicist Gaston Planté invented the first rechargeable battery. This secondary battery was based on *lead–acid (LA) chemistry*, a system that is still used today. In 1899 the Swedish Waldmar Jungner invented the *nickel–cadmium (NiCd) battery*, based on nickel for the positive and cadmium for the negative electrode. Two years later, Edison came up with an alternative design by replacing cadmium with iron. Due to high material costs relative to dry cells or LA storage batteries, the practical applications of nickel–cadmium and nickel–iron batteries were limited. In 1932 Schlecht and Ackermann invented the sintered pole plate with which great improvements were achieved. These advancements were reflected in higher load currents and improved longevity. The sealed nickel–cadmium battery, as we know it today, only became available in 1947, when Neumann succeeded in completely sealing the cell [2].

Soon after the discovery, in the late 1960s, that intermetallic compounds, such as $SmCo_5$ and $LaNi_5$, were able to absorb and also desorb large amounts of hydrogen [3], it was realized that electrodes made of these materials could serve as a new electrochemical storage medium [4], [5]. In the following years the hydride-forming electrode proved to be a serious alternative to the cadmium electrode, which was widely employed in rechargeable nickel–cadmium batteries. In particular, the higher energy storage capacity, good rate capability and non-toxic properties of the chemical elements of which these hydride-forming materials were composed were great advantages in relation to the cadmium electrode [6]. The *nickel–metal hydride (NiMH) battery* became commercially available in the 1990s [7].

The first non-rechargeable *lithium batteries* appeared in the early 1970s. Attempts to develop rechargeable lithium batteries followed in the 1980s but failed due to safety problems. Because of the inherent instability of lithium metal, especially during charging, research shifted to intercalate lithium ions in host materials in Li-ion batteries. Although lower in energy density than lithium metal, lithium ion is safe, provided certain precautions are taken when charging and discharging, implemented by means of a proper charging algorithm and a safety IC in series with the battery as discussed in the previous chapter. In 1991, the Sony Corporation commercialised the first *lithium-ion battery (Li-ion)* [1].

Table 2.1 summarises the history of the battery developments described above. The general operational mechanism of a battery and characteristics of the most important rechargeable batteries available on the market today, *e.g. nickel–cadmium, nickel–metal hydride* and *lithium-ion batteries*, will be given in the remainder of this section.

Table 2.1. History of battery development [2].

Year	Researcher (Country)	Method
1800	Volta	Invention of the battery
1859	Plante (France)	Invention of the lead-acid battery
1899	Jungner (Sweden)	Invention of the nickel-cadmium battery
1901	Edison (USA)	Invention of the nickel-iron battery
1932	Schlecht & Ackermann (Germany)	Invention of the sintered pole plate
1947	Neumann (France)	Successful sealing of the nickel-cadmium battery
1990	Sanyo (Japan)	First commercial introduction of the NiMH battery
1991	Sony (Japan)	First commercial introduction of the Li-ion battery

2.2.1 General operational mechanism of batteries

In its simplest definition, a battery is a device capable of converting chemical energy into electrical energy and *vice versa*. The chemical energy is stored in the electroactive species of the two electrodes inside the battery. The conversions occur through electrochemical reduction-oxidation (redox) or charge-transfer reactions [1]. These reactions involve the exchange of electrons between electroactive species in the two electrodes through an electrical circuit external to the battery. The reactions take place at the electrode/electrolyte interfaces. When current flows through the battery, an oxidation reaction will take place at the anode and a reduction reaction at the cathode. The oxidation reaction yields electrons to the external circuit, while a reduction reaction takes up these electrons from the external circuit. The electrolyte serves as an intermediate between the electrodes. It offers a medium for the transfer of ions. Hence, current flow is supported by electrons inside the electrodes and by ions inside the electrolyte. Externally, the current flows through the charger or load [1]. The basic electrochemical unit of a battery is called a *cell*, but the word *battery* is commonly used for one cell or for two or more cells connected in series/parallel.

During a battery's lifetime, its performance or 'health' tends to deteriorate gradually due to irreversible physical and chemical changes that take place with usage and with age until the battery is finally no longer usable. The State-of-Health (SoH) is an indication of the point that has been reached in a battery's life cycle and a measure of its condition relative to that of a fresh battery. Aging of the battery is a complex process that involves many battery parameters (*e.g.* impedance, capacity), the most important of which is capacity. To illustrate the phenomena, Fig. 2.1 shows the discharge capacity (Q_d) of a Li-ion battery represented as a function of the cycle number (C_n). The degradation curve has a clearly visible transfer point at which the rate of the battery's degradation increases.

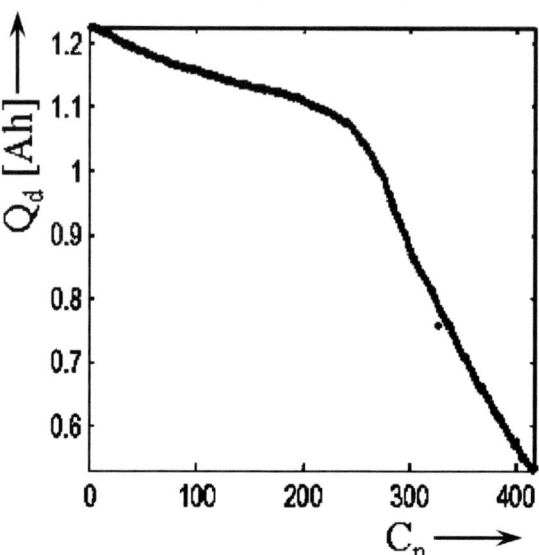

Fig. 2.1. Decrease of the discharge capacity Q_d [mAh] as a function of the operational conditions. The horizontal axis shows the cycle number C_n.

The exact position of the transfer point varies, depending on the type of battery and operational conditions. Aging of Li-ion batteries is not an absolutely new topic in modern electrochemistry. A number of models describing the aging effects in Li-ion batteries have recently been introduced. Models describing the dynamics of lithium consumption in Li-ion batteries are discussed by Broussely and Spotnitz in [8], [9]. Both Broussely and Spotnitz deal with the aging of the batteries under 'on float' condition, with the battery being kept under constant voltage of fixed polarity. A striking feature of all plots in the aforementioned articles is that they are smooth and do not have any transfer points.

A first conclusion is that it will be difficult to take into account every nuance of a battery's charge and discharge qualities and aging characteristics in an SoC indication system.

2.2.2 Battery types and characteristics

NiCd batteries. Batteries based on a positive nickel electrode of a $Ni(OH)_2/NiOOH$ compound and a negative cadmium electrode of Cd and $Cd(OH)_2$ are called NiCd batteries. The electrolyte is an aqueous KOH solution. A great advantage of NiCd batteries is their fast charge and discharge performance: it is possible to charge a battery in 10 minutes and large currents can be supplied during discharge. NiCd batteries have an average operating voltage of 1.2 V and can be used in many devices. NiCd batteries are used especially in tools demanding a lot of power. Other applications include cordless and mobile phones, shavers, camcorders, portable audio products and laptop computers. Disadvantages of NiCd batteries are their relatively low energy density and their so-called memory effect. This memory effect causes the battery to deliver only the capacity used during the preceding repeated charge/discharge cycles. Because of this effect the entire capacity of NiCd batteries should preferably be used for each discharge cycle to avoid a decrease in the maximum capacity [1]. Another disadvantage is the presence of cadmium, which is an environmental hazard. This will lead to a complete ban on NiCd batteries in the future.

NiMH batteries. The main difference between NiCd and NiMH batteries is the fact that in NiMH batteries a metal hydride alloy is used for the negative electrode instead of cadmium. In this way a higher energy density is obtained and the memory effect and environmental impact are reduced. NiMH batteries can moreover replace NiCd batteries because they have the same 1.2 V average operating voltage per cell. Applications include cordless and mobile phones, shavers, camcorders, portable audio products, laptop computers and Hybrid Electrical Vehicles (HEVs). A disadvantage of NiMH batteries is their relatively high self-discharge rate and relatively poor robustness with respect to overcharging, which is made worse by the fact that it is more difficult to detect the battery-full condition during charging [1].

Li-ion batteries. A schematic representation of a typical Li-ion cell is shown in Fig. 2.2 [10]. The cell consists of five regions (from left to right in Fig. 2.2): a negative-electrode current collector made of copper, a porous composite negative insertion electrode, a porous separator, a porous composite positive insertion electrode and a positive-electrode current collector made of aluminium. The composite electrodes are made of active material particles held together by a binder and a suitable filler material such as carbon black. When discharge is about to begin the negative electrode is fully lithiated and the positive electrode is ready to accept lithium ions. During discharge, the lithium ions deintercalate from the negative electrode particles and enter the solution phase, while in the positive electrode region lithium ions in the solution phase intercalate into the $LiCoO_2$ particles. This results in a concentration gradient, which drives lithium ions from the negative electrode to the positive electrode. The cell voltage decreases during discharge, as the equilibrium potentials and overpotentials of the two electrodes are strong functions of the concentrations of lithium on the surface of the electrode particles. The cell is considered to have reached the end of discharge when its voltage drops to 3.0 V [10].

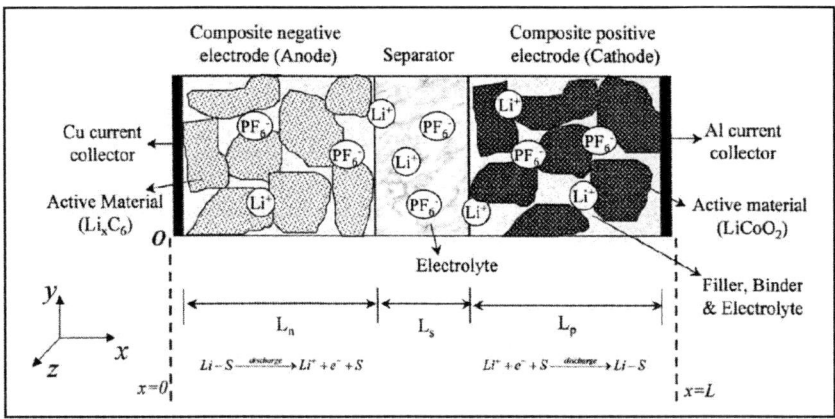

Fig. 2.2. A schematic representation of a typical Li-ion cell [10].

The positive electrode is made of lithium metal oxides (*e.g.* $LiCoO_2$, $LiNiO_2$ or $LiMn_2O_4$) for storing the lithium ions. The negative carbon electrode is made of graphite or petroleum coke. The electrolyte is usually a salt dissolved in an organic solvent, but batteries with other solvents such as propylene carbonate also exist. An example of an employed salt is $LiPF_6$. The operating voltage of the Li-ion batteries is critical and over(dis)charging results in fast aging and may cause fire or even exploding batteries. As discussed in chapter 1, an essential electronic protection circuit is consequently required to prevent over(dis)charging. Applications include mobile phones, shavers, camcorders, portable audio products and laptop computers.

2.2.3 Summary

Several types of rechargeable battery systems have been discussed in this section. The main characteristics of the discussed battery types are summarised in Table 2.2 [1].

Table 2.2. Overview of the main characteristics of the most important rechargeable battery systems.

Battery system	NiCd	NiMH	Li-ion
Average operating voltage (V)	1.2	1.2	3.6
Energy density (Wh/l)	90 – 150	160 – 310	200 – 280
Specific energy (Wh/Kg)	30 – 60	50 – 90	90 – 115
Self-discharge rate (%/month) at 20°C	10 – 20	20 – 30	1 – 10
Cycle life	300 – 700	300 – 600	500 – 1000
Temperature range (°C)	–20 – 50	–20 – 50	–20 – 50

2.3 History of State-of-Charge indication

In this section, previous and current SoC technologies will be presented. For almost as long as rechargeable batteries have existed, systems capable of indicating the amount of charge available inside a battery have been around. In 1938 Heyer introduced a single-meter device on which the value of a storage battery capacity is indicated [11]. The battery capacity is indicated on the basis of the measured battery voltage and a measured voltage drop across a sense resistor. When the battery is fully charged the device indicates 100% capacity (see Fig. 2.3). A simple test is done to determine when the battery should be replaced. In this test the voltage drop across a sense resistor during discharge from 100% capacity is measured. It is assumed that the voltage drop will be small in a fresh battery and high in an aged battery, implying that the battery should be replaced. Replacement is required when the battery capacity falls below 70% in the discharge test (see Fig. 2.3).

In 1963 Curtis Instruments pioneered gauges for monitoring the SoC, the 'fuel' level, of vehicle traction batteries. One of the methods used by Curtis involves predicting a battery's remaining capacity by measuring the amount of time elapsed since the loaded voltage dropped below a certain value. For example, when a battery is discharged from 24.25 V for a period of 3 minutes, the remaining capacity falls from 100% to 90%, *etc.* [12]. A compensation method for different discharge rates is also presented.

State-of-the-Art of battery State-of-Charge determination 17

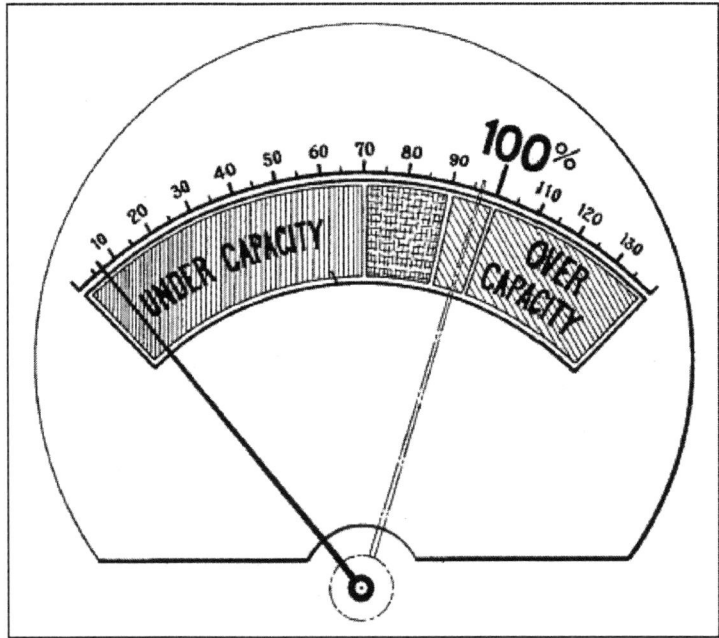

Fig. 2.3. Battery capacity indicator developed by Heyer (1938) [11].

Several monitors based on the average voltage were available in those days, such as Sears battery monitor, a range indicator (by Motovator) and the aforementioned Curtis fuel gauge. Of these, the Curtis fuel gauge was found to be the most sophisticated and accurate [13]. Curtis SoC gauges were even used on the Moon (see Fig. 2.4) [14].

Fig. 2.4. Astronauts exploring the Moon in Lunar Roving Vehicles in 1971–1972 relied on early Curtis gauges [14].

In one of the Curtis patents filed by Finger *et al.* in 1975 the current flowing from the battery is sent to an integrator module, which registers the current depletion [15]. During charging, the current is integrated in the integrator module providing a continuous display of the SoC and information needed to regulate the charge rate.

An account of the attempts made to develop an SoC indicator for a nickel–cadmium battery was given by Lerner in 1970 [16]. He concluded that the only reliable way of estimating the SoC is to use a current-sharing method. In this method, the current output of a battery having a known SoC is compared with that of a battery having an unknown SoC. The SoC of the unknown battery can be deduced from the outcome of this comparison.

In 1974 York *et al.* introduced an SoC indicator in which the value of the measured battery voltage is indicated with respect to two voltage levels stored in the system [17]. In a first state it is indicated that the battery voltage is greater than a first voltage level and in a second state it is indicated that the battery voltage is less than the first voltage level but greater than a second voltage level. Finally, a third state indicates that the battery voltage is less than the second voltage level and disables the load connected to the battery. The magnitude and duration of the voltage reduction are monitored by a threshold circuit, which produces an output whenever the terminal voltage falls below the lower threshold value. As a response to the voltage reduction, a number of pulses are generated. An electronic counter is used for counting the pulses and accumulating the counts. An integral proportional to the total time that the terminal voltage is below the lower threshold voltage is generated. The output of the integration provides an indication of the SoC. A principal advantage of this method is that SoC can be indicated despite sudden disconnection and reconnection of a battery.

The main concept of determining a battery's SoC on the basis of a comparison between the measured battery voltage and predetermined threshold values that correspond to different SoC values is also presented in [18]–[20]. In [18] the average current consumption of a portable device is determined by using a capacitor for energy storage. The capacitor is charged and discharged between two voltage thresholds in a certain measured time. The number of charging processes occurring in the measured time are counted and used to calculate the average current consumption. The maximum error obtained for the average current calculation using the above method was smaller than 5% [18]. In [19] it is shown that the approximation of the accumulated operating time of a rechargeable battery is a function of the internal resistance of the rechargeable battery. The SoC is measured relative to the rechargeable battery's maximum capacity. The applications presented in [18], [19] are described in the field of mobile telephony. In [20] voltage levels are measured during battery charge and discharge and compared with predetermined values, which are modified as a function of temperature. The stored battery charge or discharge curves are divided into curve portions defined by voltage levels and rates of changes of the voltage levels. Each curve portion defines a particular SoC of the battery. Measured voltage levels and rates of changes are then associated with the predetermined charge or discharge curve portion and the defined SoC.

In 1974 Brandwein *et al.* developed a device for monitoring nickel–cadmium batteries [21]. In addition to voltage measurements, the current that flows into and out of the battery and the battery temperature are measured and used in order to provide SoC indication. The battery voltage is stored as an analog voltage signal at different temperatures. [22] presents an equivalent circuit diagram of the battery that uses current, voltage and temperature measurements as inputs. The measured values are compared with the corresponding values obtained in calculations using the equivalent circuit diagram. The parameters in the equivalent circuit diagram and the state variables are varied so that the measured values are matched and the SoC can be inferred from the adjusted parameters.

In 1975 Christianson *et al.* developed a method in which a battery's SoC is indicated on the basis of the open-circuit voltage (OCV) calculation [23]. The OCV is directly proportional to the battery SoC and can be calculated using the following equation:

$$OCV = V_{bat} + IR \qquad (2.1)$$

where V_{bat} is the battery terminal voltage, I the actual battery current – regarded as a positive value during discharge and a negative value during charge – and R is the internal resistance. Note that $OCV = V_{bat}$ when $I = 0$, but after current interruption this takes a while due to several relaxation processes occurring inside a battery. In addition to the OCV, the method presented by Eby *et al.* in 1978 also uses the voltage under load to determine the SoC of an LA storage battery during a discharge cycle [24]. The battery's initial OCV is stored in a settable memory. It has been demonstrated that in the case of LA batteries the OCV has a linear correlation to the charge level of the battery under a defined set of circumstances. The discharge rate can be determined at any moment in time as a comparison function between voltage under load and the corresponding OCV.

The first impedance measurements of batteries appear to have been made by Willihnganz in 1941 [25]. They involved excitation of the electrochemical cell by an ac voltage of small amplitude of about 5 mV and evaluation of the resistive and reactive components or other related parameters such as the modulus of impedance and phase angle. As such measurements encompass a wide range of ac signal frequencies, various characteristic parameters of the electrochemical cell and the kinetics of the associated reactions can be evaluated [25]. As an alternative, Dowgiallo *et al.* (1975) and Zaugg (1982) developed methods for determining a battery's SoC on the basis of impedance measurements [26], [27]. The phase angle between the ac voltage across the battery terminals and the ac current through the battery (measured as a voltage through a sense resistor) is continuously monitored. The method presented in [26] relates to nickel–cadmium batteries and can be used in equipment such as transmitters, receivers, tape recorders, movie cameras, aircraft, electric vehicles, small calculators and computers. In the system developed by Muramatsu in 1985 the relationships between the battery impedance at different frequencies (defined as impedance spectroscopy), remaining capacity and SoH are used to detect a battery's SoC and SoH [28]. Predetermined values based on this relationship are stored in look-up tables and used to determine the battery's SoC and SoH. Look-up tables are tables in which fixed values of measured parameters, such as voltage, current, impedance and temperature, can be stored and used in order to indicate the SoC.

In 1984 Peled developed a method for determining the SoC of lithium-ion batteries [29]. The presented method is based on predetermined voltage and temperature measurements that are used as input parameters for look-up tables. After a current step and a short resting period, a battery's OCV and temperature are measured. The measured value is compared with a corresponding predetermined value stored in a look-up table. The outcome of this comparison is used to indicate the SoC. In the system developed by Kopmann (1987) the terminal voltage, the current and time are measured during each battery charging and discharging cycle [30]. These values are also used as inputs for look-up tables. The characteristic of the terminal voltage curve during charging and discharging is used to minimise the differences between the measured values and the battery's actual SoC. [31] presents a method for determining the SoC of NiMH batteries in notebook applications. The method uses the battery's temperature, voltage and discharge/charge rate measurements to determine the SoC, using look-up tables. A possible hardware and software implementation is also presented. These techniques can be employed for any other battery technology by modifying the look-up tables.

In 1981 Finger of Curtis Instruments patented a method according to which the SoC of Lead-Acid (LA) batteries is determined during a quiescent interval with no current flowing through the battery [32]. The battery terminal voltage is measured after a current step and the combination of these two measurements (battery voltage and time) is used for battery OCV recovery characteristics determination. This predictable time function of voltage recovery is substantially independent of the actual voltage level of the terminal voltage.

[33] presents an SoC indicator for a lithium-ion battery based on a cell OCV. The SoC is calculated on the basis of a stored SoC–OCV relationship. The SoC–OCV relationship is obtained by defining the battery charge amount when the OCV of the cell is 3.9 V as SoC = 100% and defining the battery charge amount when the OCV of the cell is 3.5 V as SoC = 0%. By defining SoC = 100% and SoC = 0% the SoC can be correctly calculated and displayed even when a battery ages.

The methods presented in [34]–[38] use Coulomb counting, *i.e.* battery current measurement and integration, as a basis. The method developed by Aylor (1992) holds for LA batteries [34]. The described technique is a combination of the previously described OCV method and coulometric measurements (Coulomb counting). The paper states that it is possible to compensate for the weakness of both techniques and provide an accurate SoC indication by combining these two measurements. Coulometric measurements are used in short time operations, in which the accumulation of error is negligible. The error that is accumulated in coulometric measurement techniques can be corrected by taking an OCV reading every time the battery has rested sufficiently. A method for predicting the OCV before the battery voltage has fully stabilised has been developed in order to reduce the required rest period of the OCV measurements. It should be noted that this method is limited to LA-type batteries because of the unique linear relation between the OCV and the specific gravity that exists in LA batteries. In [39] it is indicated that the method presented in [34] could provide 99% accuracy (1% error) for charge detection, but at a high cost of realization. The application described in [37] relates to a battery pack using NiMH batteries or any other battery technology such as LA, lithium polymer, *etc.*, operating in a hybrid–electric power train for a vehicle. Besides Coulomb counting, the systems presented by Kikuoka [35] and Seyfang [36] also compensate for temperature, charging efficiency of the battery, self-discharge and aging. In [36] a battery's capacity is monitored and compared with the initial capacity to obtain an indication of the battery's SoH. When the battery is fully discharged or fully charged a particular set of parameters, such as the conversion efficiency of each battery, are 'learned' and updated to take into account the battery's aging. Besides Coulomb counting and voltage, current and temperature measurements, [38] presents a mathematical model that is implemented on a computer that simulates a battery's behaviour.

The methods presented in [40]–[42] also use adaptive methods for determining a battery's SoC. In 1997 Gerard *et al.* developed a method in which a battery's 'state variables' are replaced with neural weights with the aim of providing portable equipment users with an accurate estimation of the remaining working time, *e.g.* how much time is left until the battery voltage reaches the end-of-discharge voltage defined in a portable device [40]. Two artificial neural networks are used to model the system's implementation, more precisely, to adapt the prediction of the current discharge curve to the general behaviour of the employed battery pack. A mean error of about 3% has been found using this implementation method. In 1999 Salkind *et al.* developed a method for SoC and SoH prediction based on fuzzy logic modelling for two battery systems – lithium–sulphur dioxide ($Li-SO_2$) and NiMH [41]. The method involves the use of fuzzy logic mathematics to analyse data obtained by impedance spectroscopy and/or Coulomb counting techniques. The maximum error between the measured SoC and the model-predicted SoC obtained using the above method for a limited data set in the case of lithium–sulphur dioxide cells was +/– 5%. In 2000 Garche *et al.* developed a method in which the Kalman filters (KF) are used to implement an adaptive method in connection with parameter estimation to determine SoC [42]. The basis of the filter is a numeric battery model description. The battery voltage is estimated on the basis of the current and temperature measurements and the results are compared with the measured battery voltage value. Adaptivity of the model is based on a comparison of the estimated values with observed battery behaviour. A more detailed description of these adaptive methods will be given in section 2.5.

In 2000 Bergveld *et al.* developed a method for estimating the SoC of a rechargeable lithium-ion battery [1], [43]. The basis of the algorithm is current measurement during the charge or discharge state and voltage measurement during the equilibrium state (state in which no current is flowing into or out of the battery and all the conditions inside the battery are fully stabilised). In the charge and discharge states the determination of the SoC relies on calculating the charge withdrawn from or supplied to the battery by means of current integration and subtracting this charge from or adding it to the previously calculated SoC. So in these states Coulomb counting is applied and the battery is viewed as a simple linear capacitor.

In addition to simple Coulomb counting the effect of the overpotential is also considered in the discharge state. Due to this overpotential, the battery voltage during discharging is lower than the Electro-Motive Force (EMF equals the sum of the equilibrium potentials of a battery's electrodes), which in equilibrium equals the OCV presented above. The value of the overpotential depends on the discharge current, the SoC, age and temperature. Especially at low temperatures and low SoC values the remaining charge cannot be withdrawn from the battery due to a high overpotential caused mainly by diffusion limitation of the electrochemical reactions, because otherwise the battery voltage will drop below the end-of-discharge voltage defined in the portable device. This leads to an apparent capacity loss, which at low temperatures of *e.g.* 0°C may amount to more than 5%. So a distinction should be made between the charge that is available in the battery and the charge that can be withdrawn from the battery under certain conditions. As overpotentials are temperature dependent, temperature measurements must also be carried out in the discharge state [1].

In the equilibrium state, a battery's SoC is determined by means of voltage measurements. Because only a negligibly small current flows in this state, the measured voltage approaches the battery's EMF. The algorithm uses a stored EMF versus-SoC curve to translate a measured voltage value into an SoC value expressed in % of the maximum capacity. The EMF-versus-SoC curve remains the same even when the battery ages and the temperature dependence of this curve is relatively low [1], [44]. The EMF method can also be used to calibrate the SoC system because the same SoC has been found for a certain measured EMF, irrespective of the battery's age and temperature. This calibration is important, because in the charge and discharge states the calculated SoC will eventually drift away from the real value due to *e.g.* measurement inaccuracy in the current and the integration in time of this inaccuracy [1]. A complete description of this algorithm, which is also the starting point of this book, will be given in the next chapter.

Table 2.3 summarises the most important points of the history of SoC development outlined above.

Table 2.3. History of SoC development.

Year	Researcher/ Company	Method
1938	Heyer	Voltage measurements
1963	Curtis	Voltage measurements and threshold in voltage levels
1970	Lerner	Comparison between two batteries (one with a known SoC)
1974	Brandwein	Voltage, temperature and current measurements
1975	Christianson	OCV
1975	Dowgiallo	Impedance measurements
1975	Finger	Coulomb counting
1978	Eby	OCV and voltage under load
1980	Kikuoka	Book-keeping
1981	Finger	Voltage relaxation
1984	Peled	Look-up tables based on OCV and T measurements
1985	Muramatsu	Impedance spectroscopy
1986	Kopmann	Look-up tables based on V, I and T measurements
1988	Seyfang	Book-keeping and adaptive system
1992	Aylor	OCV, OCV prediction and coulometric measurements
1997	Gerard	Voltage and Current Measurements, Artificial Neural Networks
1999	Salkind	Coulomb counting, impedance spectroscopy, fuzzy logic
2000	Garche	Voltage and Current Measurements, Kalman filters
2000	Bergveld	Book-keeping, overpotential, EMF, maximum capacity learning algorithm

As will be shown later in this chapter, the actual State-of-Charge indication integrated circuits (ICs) are based on the methods indicated in this table. More detailed information on the applied methods will be given in section 2.5.

2.4 A general State-of-Charge system

In science, the standard unit used to express battery capacity is Coulomb (named after the French physicist C. A. Coulomb, 1736–1806), which describes the time a battery can produce a given current. The Coulomb is the unit of electric charge corresponding to one ampere-second (As). In practice, however, cell or battery capacity is more commonly expressed in ampere-hours (Ah) or milliampere-hours (mAh). Of great importance for users is to know a battery's SoC. In [37] SoC is defined as the percentage of the full capacity of a battery that is still available for further discharge. In [25] it is the ratio of a cell's available capacity and its maximum attainable capacity. For a proper understanding of what the term 'SoC' really implies a clear definition is needed: SoC is the percentage of maximum possible charge that is present inside a rechargeable battery. The SoC measurement method and the computational model based on the correct SoC definition must be simple, convenient, practical and reliable.

Fig. 2.5 shows an example of a practical SoC system. The battery may include a plurality of battery cells connected in series and/or parallel, each of the battery cells having at least two terminals. The SoC system may include an analogue-to-digital converter (ADC) for converting a voltage drop between at least two sense resistor connection pins as a measure of the current *(I)* into a digital signal and also for converting the measured analogue values of the battery voltage *(V)* and temperature *(T)* into digital signals.

A microprocessor/microcontroller (in which the SoC algorithm is stored) determines a battery system's SoC on the basis of the measured signals. Two types of memory are needed. Basic battery data, such as the amount of self-discharge as a function of *T* and the discharging efficiency as a function of *I* and *T*, are read from the read-only memory (ROM). When the SoC algorithm is based on EMF measurements, the EMF–SoC relationship can be stored in ROM together with other battery-specific data. The random access memory (RAM) is used to store the history of use, such as the number of charge/discharge cycles, which can be used to update the maximum battery capacity. Each part of this system (software algorithm or hardware device) will influence the ultimate accuracy of the SoC indication (*e.g.* inaccuracy in the *V*, *T* and *I* measurements will result in inaccuracy in the final SoC). Also important is the calibration of the SoC, because if the SoC algorithm is based on, say, current measurement and integration, the error caused by the current measurement inaccuracy will accumulate over time.

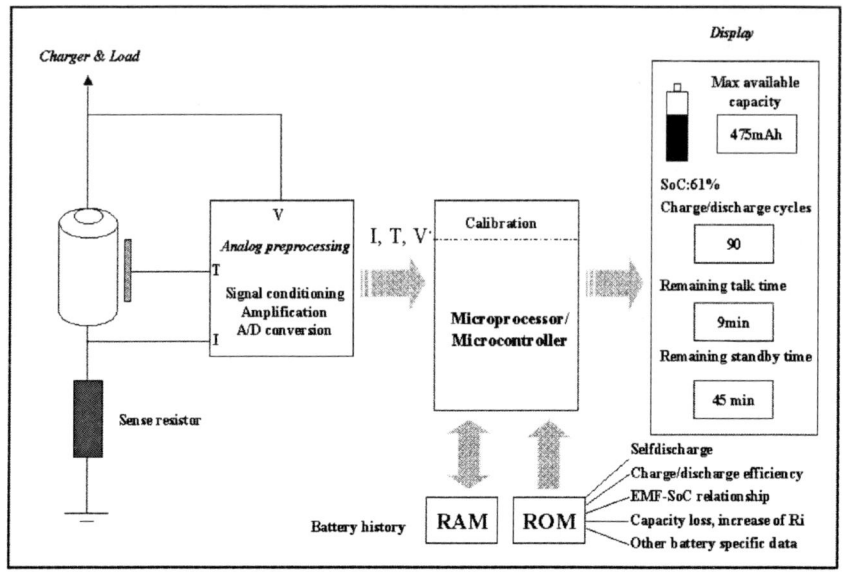

Fig. 2.5. General functional architecture of a State-of-Charge system.

2.5 Possible State-of-Charge indication methods

For an efficient discussion of SoC indication methods, some of the terms commonly used in the battery SoC industry should be defined.

Ampere-hour. A measure of electric charge defined as the integral product of current (in Amperes) and time (in hours).

Cell. The basic electrochemical unit used to generate electrical energy from stored chemical energy or to store electrical energy in the form of chemical energy. A cell consists of two electrodes in a container filled with an electrolyte.

Battery. Two or more cells connected in an appropriate series/parallel arrangement to obtain the operating voltage and capacity required for a certain load. The term is also frequently used for single cells.

Li-ion cells. Cells containing a liquid organic or polymer electrolyte in which the anode and cathode are both made of intercalation compounds [46].

C-rate. A charge or discharge current equal in Amperes to the rated capacity in Ah. Multiples larger or smaller than the C-rate are used to express larger or smaller currents. For example, the C-rate is 1100 mA in the case of an 1100 mAh battery, while the C/2 and 2C-rates are 550 mA and 2.2 A, respectively.

Capacity. A battery's electrical energy content expressed in ampere-hours.

Maximum capacity. Maximum amount of capacity that can be removed from a battery under defined discharge conditions.

Cycle life. The number of cycles that a cell or battery can be charged and discharged under specific conditions before the available capacity in Ah fails to meet specific performance criteria. This will usually be 80% of the rated capacity.

Cut-off voltage. The lowest operating voltage at which a cell is considered depleted. Also often referred to as end-of-discharge voltage or final voltage [34].

Self-discharge. The recoverable loss of a cell's useful capacity on storage due to internal chemical action. This is usually expressed in a percentage of the rated capacity lost per month at a certain temperature because batteries' self-discharge rates are strongly temperature-dependent. The self-discharge mechanism is a local redox process caused by decomposition of the electrolyte [46]. Other important sources for the self-discharge are micro-shorts and shuttle-molecules.

Spread. Difference between characteristics of batteries of the same type.

State-of-Health (SoH). A 'measurement' that reflects a battery's general condition and its ability to deliver the specified performance in comparison with a fresh battery.

State-of-Charge (SoC). The percentage of the maximum possible charge that is present inside a rechargeable battery.

Depth-of-Discharge (DoD). The amount of capacity withdrawn from a battery expressed as a percentage of its maximum capacity.

Depth-of-Charge (DoC). The amount of capacity put into a battery expressed as a percentage of its maximum capacity.

Remaining run-time. The estimated time that a battery can supply current to a portable device under valid discharge conditions before it will stop functioning.

As indicated in section 2.3, there are several methods for determining the SoC of a battery. Some early, very inexpensive fuel gauges simply measured voltage. Battery voltage is a highly inaccurate indication of a battery's capacity because it changes with temperature, discharge rates and aging. Another known method for measuring SoC involves impedance measurements. The measurements obtained are compared with previously generated standard reference curves. Yet another prior-art method used to determine battery SoCs involves estimating the SoC on the basis of a battery's response to current or voltage pulses. These pulse systems yield only a very general impression of a SoC and are used primarily to determine whether a battery is still useable. This first group of methods will be called *direct measurements* below.

Another known method is to measure the current flowing into and out of a battery and to integrate this current over time in order to determine its capacity [1]. When using these current integrators one must correct the estimation of the SoC obtained because several battery-related factors affect the accuracy of the estimation. These factors include temperature, history, charge and discharge efficiencies and cycle life. The integration of current is referred to in the literature as *Coulomb counting* [1]. When discharging 'efficiency', self-discharge and capacity loss are compensated for, this method can be regarded as a *book-keeping system* [1].

The main problem in designing an accurate SoC indication system is the unpredictability of the behaviour of both batteries and users. For this reason use must be made of an adaptive system based on direct measurement, book-keeping or a combination of the two [1]. In order to clarify all aspects, these methods will be discussed separately below.

2.5.1 Direct measurement

The direct measurement method refers to the measurement of battery variables such as the battery voltage (V), battery impedance (Z) and voltage relaxation time (τ) after application of a current step. Most relations between battery variables and the SoC depend on the temperature (T). This means that the battery temperature should also be measured, besides the voltage or impedance. The basic principle of a SoC indication system based on direct measurement is shown in Fig. 2.6.

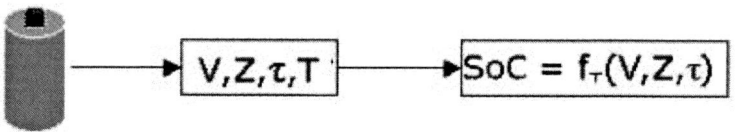

Fig. 2.6 Basic principle of a SoC indication system based on direct measurement.

The main advantage of a system based on direct measurement is that it does not have to be continuously connected to the battery. The measurements can be performed as soon as the battery has been connected [1].

Voltage measurements. Although voltage measurement has been a popular method, especially for mobile phone applications, it does not produce the most accurate results. Determining the remaining capacity of a cell simply by measuring its voltage level may be less expensive and may use less computing power of the host CPU than Coulomb counting, but under real-life conditions voltage measurements alone can be very misleading [47]. While it is true that a given cell voltage level will continually drop during discharge, the voltage level in relation to remaining charge varies greatly with cell temperature and discharge rate. Fig. 2.7 shows an Li-ion battery voltage curve during discharge at different discharge rates.

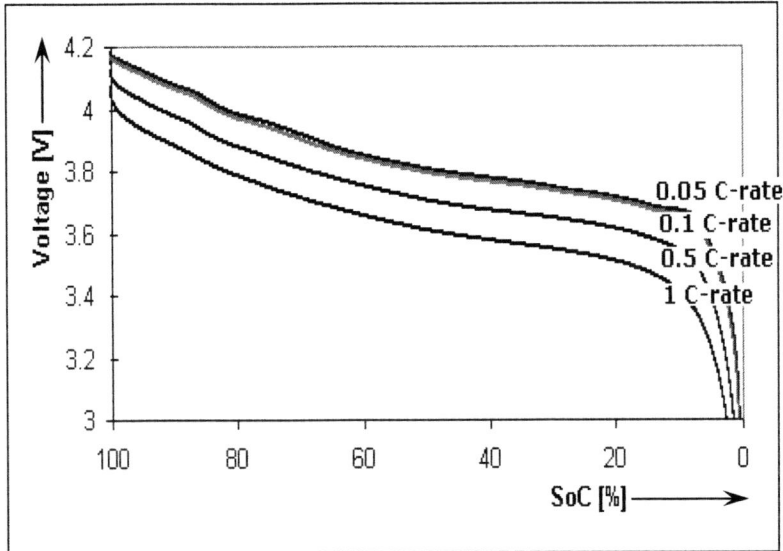

Fig. 2.7. Li-ion battery voltage curves at different discharge rates.

The important point illustrated in Fig. 2.7 is the relationship between the cell voltage and its discharged capacity. It can be seen that the voltage discharge curve depends strongly on the discharge rate. The error in SoC estimation based on voltage measurement can be corrected by the system if the dependence of the battery voltage on the cell temperature and discharge rate is known. However, when those measured curves are included the process becomes more complicated and expensive than a Coulomb counting approach [47].

The EMF method. The term EMF stands for electromotive force. It is a battery's internal driving force for providing energy to a load. In principle, the EMF can be inferred from thermodynamic data and the Nernst equation (or equations derived from it) [1]. Another method with which the EMF can be obtained is called *linear interpolation*. With this method the average battery voltage, calculated at the same SoC, is inferred from the battery voltages during two consecutive discharge and charge cycles using the same currents and at the same temperature. Fig. 2.8 shows the EMF curve obtained at 25°C with the linear interpolation method using Sony's US18500G3 Li-ion battery.

In another known method the EMF is determined on the basis of *voltage relaxation*. The battery voltage will relax to the EMF value after current interruption. This may take a long time, especially when a battery is almost empty, at low temperatures and after a high discharge current rate [44]. In another EMF determination method – *linear extrapolation* – the battery voltages obtained with different currents with the same sign and at the same SoC value are linearly extrapolated to a current of value zero [1].

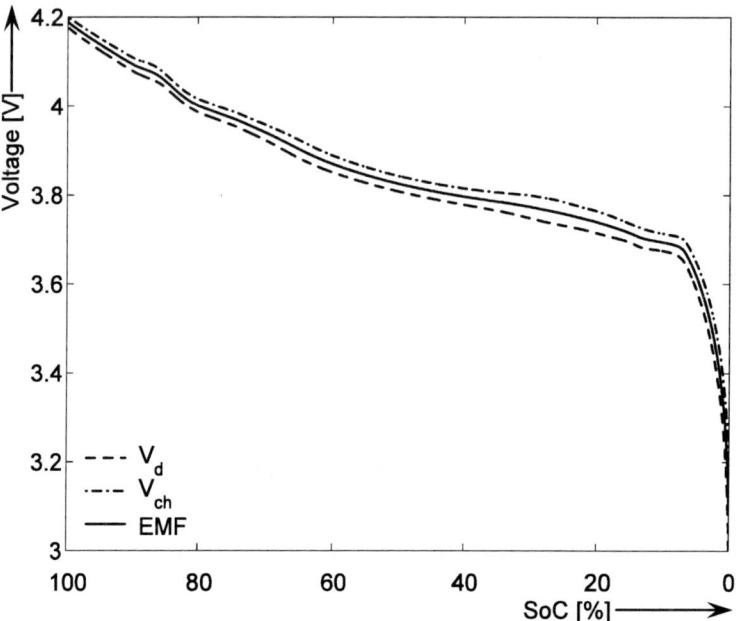

Fig. 2.8. EMF curve obtained with the linear interpolation method.

The EMF of an Li-ion battery has been found to be a good measure of battery SoC. It has been demonstrated that the relationship between the EMF and the SoC does not change during cycling of the battery if the SoC is expressed in relative capacity [1]. The temperature dependence of the EMF is small, except when the battery is almost fully discharged or almost fully charged.

When the SoC algorithm is based on the EMF an accurate method for EMF implementation is required. Three of the EMF implementation methods used in practice will be presented below.

(a) *Look-up table*. A table in which fixed values of the measured parameters can be stored and used in order to indicate SoC. The size and accuracy of the look-up tables in SoC indication systems depend on the number of stored values. One of the main drawbacks of this method is that even in the case of a single type of battery it is difficult to take into account every point of the EMF curve in order to provide an accurate SoC indication system. When many measurement points are included the process becomes more complicated and expensive than other approaches, and will probably not provide any significant advantages.

(b) *Piecewise linear function*. In this method the EMF curve is approximated with piecewise linear functions. A possible example with 10 intervals is shown in Fig. 2.9 for Sony's US18500G3 Li-ion battery. The intervals in voltage and the corresponding SoC are presented in Table 2.4.

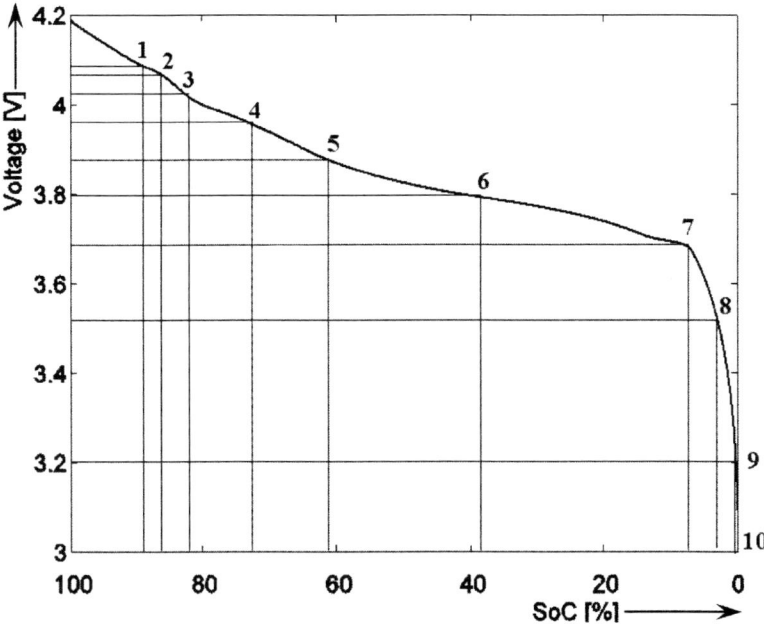

Fig. 2.9. Piecewise EMF curve for the US18500G3 Li-ion battery from Sony.

Table 2.4. Possible EMF curve implementation (see also Fig. 2.9).

Interval number	Interval voltage limits [V]	SoC [%]
1	4.08 – 4.20	90 – 100
2	4.06 – 4.08	84 – 90
3	4.02 – 4.06	81 – 84
4	3.98 – 4.02	72 – 81
5	3.88 – 3.98	61 – 72
6	3.80 – 3.88	39 – 61
7	3.68 – 3.80	8 – 39
8	3.54 – 3.68	4 – 8
9	3.22 – 3.54	0.5 – 4
10	3.00 – 3.22	0.0 – 0.5

With the aid of Eq. (2.2), the SoC for any measured battery equilibrium voltage value, *i.e.* EMF, can be calculated:

$$SoC = SoC_l + \frac{EMF - V_l}{V_h - V_l}(SoC_h - SoC_l) \qquad (2.2)$$

where V_l and V_h are fixed and specific values from the EMF curve representing the voltages corresponding to the SoC_l [%] and SoC_h [%] SoC values, *e.g.* in table 2.4 $V_l = 4.08$ V and $V_h = 4.2$ V corresponding to $SoC_l = 90\%$ and $SoC_h = 100\%$, respectively.

When enough voltage and SoC intervals are chosen, this method will allow more flexibility (possibility of implementation for other types of battery) and precision in SoC estimation based on the EMF curve in comparison with a look-up table implementation. The problems of spread, temperature and aging remain to be solved.

(c) *Mathematical function.* In this method the EMF curve is approximated with a mathematical function. A possible example in which the EMF of an Li-ion battery with intercalated electrodes will be modelled as a difference in equilibrium potentials of positive and negative electrodes will be given in chapter 4 [1], [49].

Using an adaptive method for updating the equation parameters taking into consideration factors like spread, temperature and aging of the batteries, this method will probably offer the best solution for a practical EMF implementation.

Impedance measurements. The ratio of a complex voltage and a complex current is in general a complex quantity. The ratio V/I is generally denoted as the impedance Z (see for instance [50]). This definition is not always correctly applied in the battery-related literature. Impedance data on many practical battery designs are reported in [51]. A useful way of studying processes in electrochemical systems including biological processes, batteries and capacitors is to make impedance measurements over a wide range of frequencies, usually referred to as Electrochemical Impedance Spectroscopy (EIS).

The electrochemical impedance (or ac impedance) of a battery characterises its dynamic behaviour, that is, its response to an excitation of small amplitude. In principle, any type of excitation signal may be used (sine wave, noise, step, . . .). In practice, sine waves are however usually used. In galvanostatic (constant current) mode, the dc current I (polarization current) charging or discharging the battery is modified using a sinusoidal current:

$$\Delta I = I_{max} \sin(2\pi f t) \qquad (2.3)$$

at frequency f, which is superimposed to I, yielding a sinusoidal voltage response:

$$\Delta V = V_{max} \sin(2\pi f t + \phi) \qquad (2.4)$$

around the dc voltage V at the battery's terminals. The amplitude V_{max} and the phase angle ϕ depend on the frequency f and V_{max} also depends on the amplitude I_{max} of the applied ac current. In contrast, in potentiostatic mode (constant voltage), the dc voltage V at the battery's terminals is modified using a sinusoidal voltage

$$\Delta V = V_{max} \sin(2\pi f t) \qquad (2.5)$$

at frequency f, which is superimposed onto V, yielding a sinusoidal current response:

$$\Delta I = I_{max} \sin(2\pi f t - \phi) \qquad (2.6)$$

around the dc current I flowing through the battery. In this case, the amplitude I_{max} and the phase angle ϕ depend on the frequency f and I_{max} also depends on the amplitude V_{max} of the applied ac voltage. In both cases, the impedance is defined by

$$Z(f) = \frac{V_{max}}{I_{max}} e^{j\varphi} \tag{2.7}$$

Therefore, the electrochemical impedance of a battery is a frequency-dependent complex number characterised by either its real and imaginary parts or its modulus and phase angle φ.

It should be noted that the voltage amplitude V_{max} must not exceed about 10 mV to ensure that impedance measurements are performed under linear conditions. In this case the excitation and response signals are actually sine waves and the measured impedance does not depend on the amplitude of the excitation signal. Such a condition is easily fulfilled in potentiostatic mode with V_{max} being directly imposed by the experimenter. In galvanostatic mode, I_{max} must be determined so that V_{max} is close to 10 mV at all frequencies, especially at the lowest analysed frequency at which the modulus of the battery impedance is maximum. High power ac currents (of several A) may be required for high capacity batteries whose impedance values are in the mΩ range. Impedance diagrams may be presented as a Bode plot (modulus in log scale versus frequency and phase angle versus frequency) or, more frequently, a Nyquist plot (imaginary part versus real part). In the latter case, electrochemists generally plot the negative of the imaginary part on the ordinate axis, so that the capacitive loops appear in the upper quadrants. The general shape of the Nyquist diagram of the complex electrochemical impedance of a high-capacity LA battery cell is given in Fig. 2.10.

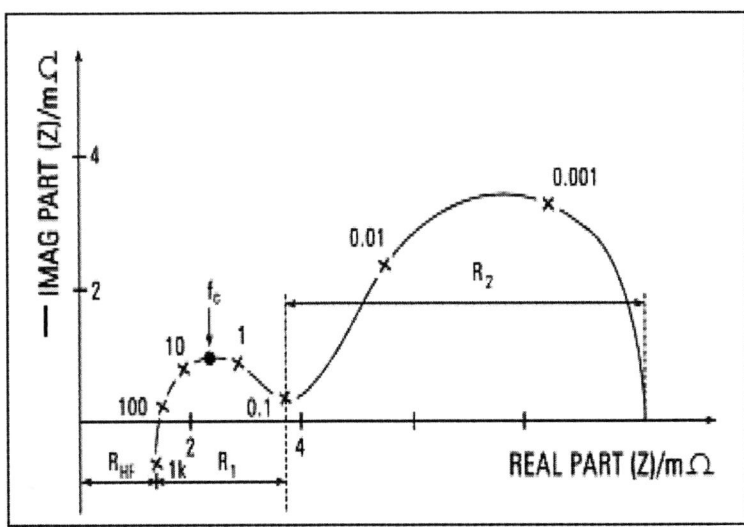

Fig. 2.10. Shape of the Nyquist diagram of the complex impedance of a high-capacity lead–acid battery cell (frequencies in Hz) [51].

The Nyquist diagram of Fig. 2.10 presents:

(a) an inductive part at frequencies higher than 100 Hz;
(b) a high-frequency resistance R_{HF} in the range of mΩ, which is the real part of the impedance at frequencies higher than 100 Hz;

(c) a first and small capacitive loop (size R_1) for frequencies between 0.1 and 100 Hz corresponding to the electrochemical reaction with the fastest kinetics;

(d) a second and large loop (size R_2) for frequencies lower than 0.1 Hz corresponding to the electrochemical reaction with the slowest kinetics.

Measurements of battery impedance as a function of frequency are not practical for SoC indication in a portable device because a signal with a frequency sweep has to be applied. Some dependence of the impedance on the SoC can be found in a laboratory set-up, but this dependence will usually be smaller than the dependence on a large temperature range as encountered in portable devices. The impedance measurements are used in portable products mainly as a means of indicating the battery's condition (SoH). Battery wear-out can be detected by an increase in the internal resistance, so the value of the internal resistance can be assessed by simply applying a current step to test whether the battery is of poor quality and should be replaced [1].

2.5.2 Book–keeping systems

Book-keeping is a method for SoC indication that is based on both current measurement and integration. This can be denoted as Coulomb counting, which literally means 'counting the charge flowing into or out of the battery'. These Coulomb counting data and other relevant data of the battery such as self-discharge rate, temperature, charge/discharge efficiency, history (*e.g.* cycle life), *etc.* will be used as input for the book-keeping system. The following Li-ion battery effects can be compensated for in a book-keeping system [1]:

Discharging 'efficiency'. Depending mainly on the SoC, T and I, only part of the available charge inside a battery can be retrieved. The main mechanisms behind this 'efficiency' are reaction kinetics and diffusion processes. These mechanisms involve reaction-rate and diffusion constants, which are temperature-dependent. Moreover, increased depletion of reacting species at the electrode surfaces occurs at larger currents and reaction-rate constants change over time as a battery ages. Consequently, a battery that may seem empty after it has been discharged with a relatively high current can still be discharged further after a rest period and/or with a lower current. In general, less charge can be obtained from a battery at low temperatures and/or large discharge currents. A battery's age also influences the discharging efficiency, for example due to increased internal resistance.

Self-discharge. Any battery will gradually loose charge, which will become apparent when a battery is left unused for some time. A Coulomb counter cannot measure this quantity of charge as no net current flows through the battery terminals. A battery's self-discharge rate will depend strongly on temperature and on the SoC.

Capacity loss. The maximum possible battery capacity in Ah decreases when a battery ages. The capacity loss depends on many factors. In general, the more the battery is misused, for example overcharged and overdischarged on a regular basis, the larger the loss will be. In most commercial book-keeping systems voltage measurement is often used to update the maximum battery capacity so as to deal with capacity loss [1].

The overall accuracy of 'Coulomb counting' depends on the accuracy of the current measurement across the full operating range of the battery system, in both the charging and discharging modes. Typically, the current measuring device measures the voltage across a shunt resistor connected in series to the battery system and converts the measured voltage into a current. This current is integrated and used to determine the SoC of the battery system. Higher current levels require substantially lower shunt resistor values and higher power dissipation ratings. The low resistance of such a shunt results in a very small voltage drop across the shunt, which must be measured in order to determine the smaller charge and discharge currents in the battery system. Since the function of the battery monitoring system is to provide time integration of the battery current in order to track the battery's SoC, even small errors in the measurement of the current can cause large errors in the SoC measurement to accumulate over time. One of the common errors when the signals to be measured are really small is the offset of the current measurement device.

A particular example of a book-keeping system for a mobile phone application is illustrated in Fig. 2.11 [52].

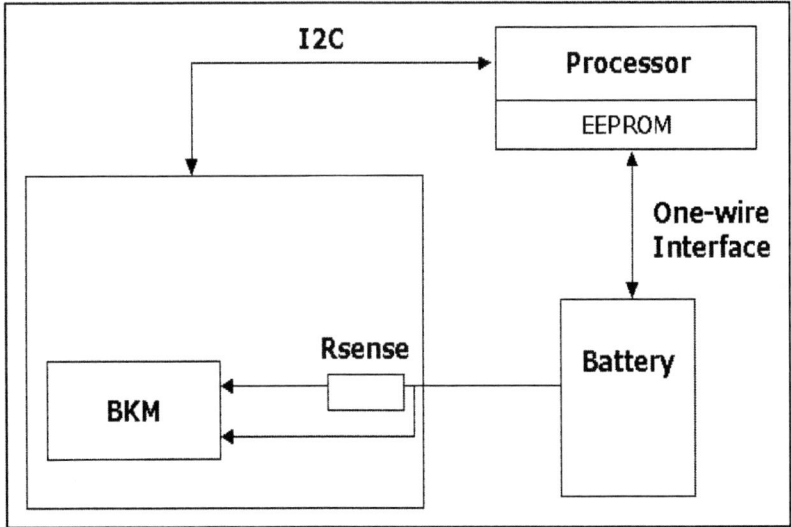

Fig. 2.11. Book-keeping system in a mobile phone application.

The book-keeping module (BKM) continuously monitors the battery and reports the obtained information (voltage, temperature, current measurements and integration) to the processor. The processor uses this information and the battery identification data to determine the SoC. The battery identification data consist of data allowing the determination of the battery's capacity. Those data are stored in the electrically erasable ROM (EEPROM) and are continuously updated by the processor. The processor uses one-wire interface to communicate with the battery. This means that the battery pack needs only three output connections: battery power, ground and one-wire interface.

The BKM can work in two different modes:

The sensitive mode. When a telephone is in idle mode, its consumption will be low, so greater accuracy will be needed to measure the current. The sensitive mode requires a measurement with high sensitivity.

The normal mode. During communication the consumption current is quite high in comparison with the idle mode. The normal mode is also used in the charge mode.

A high sensitivity, referred to as the lowest current value that needs to be measured, is very important for ensuring that all the charge flowing from and to a battery is monitored. The minor charge variations are not important for users, but they are essential for the system's reliability. The system must have some internal registers (see Fig. 2.12) that accumulate the minor variations. These registers are described below.

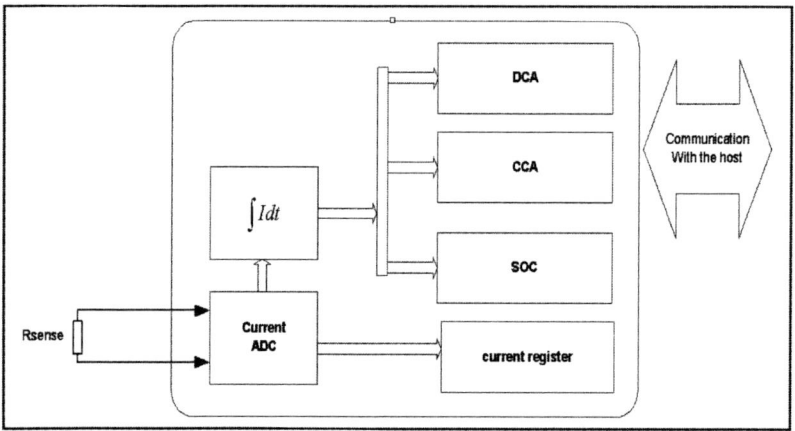

Fig. 2.12. Block diagram of the book-keeping support module (BKM).

Current register. Measurement of the current flowing into and out of a battery at the sampling moment.

SOC (SoC counter). Maintains a net accumulated charge flowing into and out of a battery. The reading in this register is an indication of the remaining capacity. This register is reset each time a low indication is given.

CCA (charging current accumulator). Accumulates the total charging current throughout a battery's life. It is only updated when the battery receives current.

DCA (discharging current accumulator). Accumulates the total discharging current throughout a battery's life. It is only updated when the battery provides current.

The DCA and CCA registers give information to the battery system needed to determine the end of life of the rechargeable battery based on total charge/discharge current throughout its lifetime.

2.5.3 Adaptive systems

The main problem in designing an accurate SoC indication system is the unpredictability of both battery behaviour and user behaviour. For this reason an adaptive system has to be used, which is based on direct measurement, book-keeping or a combination of the two [1]. Some examples of existing adaptive SoC systems will be described in this section.

In [42], [53]–[55] optimum Kalman filters are presented for implementing an adaptive method in connection with parameter estimation to determine SoC. The basis of the filter is a numeric battery model description. In [42] the battery voltage is estimated on the basis of current and temperature measurements and then the results are compared with the measured battery voltage value (see Fig. 2.13). The inner (internal) parameter contains at least the SoC, but could also contain additional battery variables, such as an estimated value of the battery series resistance (which will give information on the battery's SoH). The model may contain the direct measurement function or the book-keeping function or a combination of the two. The system starts with a basic set of information describing standard behaviour of the type of battery concerned.

Adaptivity of the model is based on a comparison of the estimated values with observed battery behaviour. This comparison is made whenever possible. The purpose of a Kalman filter is to estimate a system's state on the basis of measurements, which contain errors. The filter has the advantage of being sequential – it needs only the system variables of the previous sample and the forcing terms and observations of the current sample.

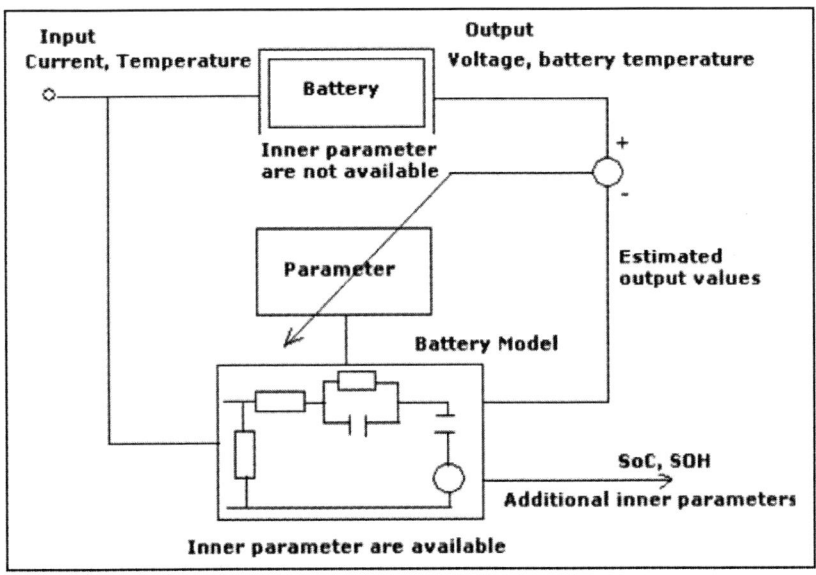

Fig. 2.13. Method for SoC and SoH determination using a Kalman filter [42].

In [53]–[55] Plett shows how the EKF (extended KF) may be used to adaptively identify unknown parameters in a cell model, in real-time, given cell voltage, current and temperature measurements. Five mathematical state-space models for modelling LiPB hybrid electrical vehicle (HEV) cell dynamics are discussed. The models with a single state are simple, but perform poorest. Adding hysteresis and filter states to the model improves performance, at some cost in complexity. The final model includes terms that describe the dynamic contributions due to open-circuit voltage, polarisation time constants, electrochemical hysteresis, ohmic loss and the effects of temperature. The results demonstrate that it is possible to achieve a root-mean-squared modelling error that is smaller than the level of quantization error expected in an implementation. It is concluded that EKF provides the best solution for long-term SoC estimation [55].

[40] presents an application in which a battery's 'state variables' are replaced with neural weights with the aim of providing the portable equipment user with an accurate estimation of the remaining working time, *i.e.* how much time is left until the battery voltage reaches the cut-off value.

Two artificial neural networks (ANN), ANN_A and ANN_O, are used to model the system's implementation, more precisely to adapt the prediction of the current discharge curve to the general behaviour of the employed battery pack (see Fig. 2.14). An ANN requires at least two phases: a training phase to set the synaptic weights at those offering the 'best compromise' and an evaluation phase to test the accuracy of the ANN using previously unseen samples. 2860 discharge curves (260 cycles for 11 batteries) were used to train the system. It has been demonstrated that the ANN (even with online adaptation) can be useful even for small products. A mean error of about 3% has been found using this implementation method.

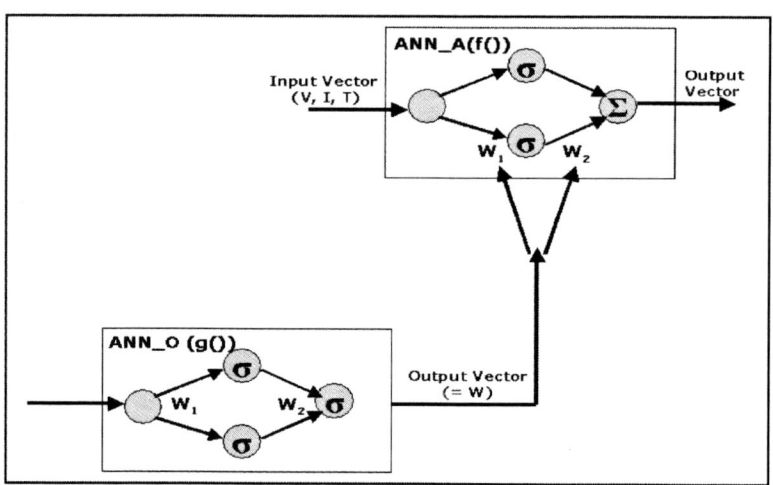

Fig. 2.14. Schematic representation of smart battery management using ANN [40].

[56] presents a method of SoC determination suitable for mobile communication applications. The effects of pulse current loads are investigated using a three-layer feed-forward artificial neural network, which is trained using the back propagation algorithm (for adjusting the weights and biases of each neuron on the basis of the error between SoC and the network's output). The paper demonstrates the use of artificial neural networks for characterising the discharge patterns for Li-ion batteries under pulsed loads typical of those required by the mobile telecommunications system.

In [41] SoC and SoH prediction based on fuzzy logic modelling is demonstrated for two battery systems, lithium–sulphur dioxide and NiMH. The method involves using fuzzy logic mathematics to analyse data obtained by impedance spectroscopy and/or Coulomb counting techniques. Data may be categorised by 'crisp' or 'fuzzy' sets. Crisp sets categorise data with certainty, *e.g.* a set of temperatures between 30°C and 40°C. With fuzzy sets, the set in which data can be categorised is uncertain, *e.g.* the temperature is 'warm'. This linguistic descriptor 'warm' is a subset of a set of all temperatures and is defined by its membership function. The degree to which an element of the 'temperature' set belongs to the fuzzy subset 'warm' is indicated by a quantity referred to as its 'degree of membership' or fit fuzzy unit value.

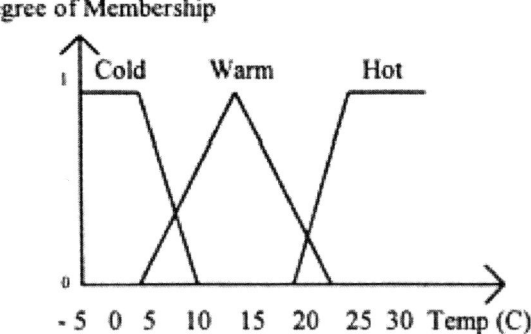

Fig. 2.15. Membership function for temperature.

Fig. 2.15 shows an example of three subsets, defined by their membership functions, 'cold', 'warm' and 'hot', of the 'universe of discourse' 'temperature' set. Using the above method for a limited data set the maximum error between the measured SoC and the model-predicted SoC obtained for lithium–sulphur dioxide cells was found to be +/– 5%.

2.5.4 Summary

A number of possible methods for SoC determination have been presented in this section. They were followed by descriptions of three of the best-known adaptive methods, which have as inputs measured battery variables such as voltage, impedance and current, and use these variables in order to accurately predict the SoC and the remaining time of use for an application. The SoC determination methods are summarised in Table 2.5 along with their fields of application, advantages and drawbacks [57].

Table 2.5. Overview of methods for SoC determination [57].

Technique	Field of application	Advantages	Drawbacks
Discharge test	Used for capacity determination at the beginning of life	Easy and accurate; independent of SoH	Offline, time-intensive, modifies the battery state, loss of energy
Coulomb counting	All battery systems, most applications	Accurate if enough re-calibration points are available and with good current measurements	Sensitive to parasite reactions; needs regular re-calibration points
OCV	Lead, Lithium, Zn/Br	Online, cheap, OCV prediction	Needs long rest time (current = 0)
EMF	Lead, Lithium	Online, cheap, EMF prediction	Needs long rest time (current = 0)
Linear model	Lead Photovoltaic	Online, easy	Needs reference data for fitting parameters
Impedance spectroscopy	All systems	Gives information on SoH and quality	Temperature sensitive, cost intensive
D. C. Internal resistance	Lead, NiCd	Gives information on SoH; possibility of online measurements	Good accuracy, but only for a short time interval
Artificial Neural Networks	All battery systems	Online	Needs training data of a similar battery, expensive to implement
Fuzzy logic	All battery systems	Online	Ask a lot of memory in real-word application
Kalman filters	All battery systems, PV, dynamic application	Online Dynamic	Difficult to implement the filtering algorithm that considers all features as *e.g.* nonnormalities and nonlinearities

2.6 Commercial State-of-Charge indication systems

The actual State-of-Charge technologies used by leading battery-management IC producers in practice will be presented in this section. Among those producers are Benchmarq/Unitrode/Texas Instruments, Dallas Semiconductor/Maxim, Linear Technology, PowerSmart/Microchip, Analog Devices, Xicor, Atmel and Mitsumi, to name but a few (see Table 2.6). Numerous manufacturers are offering SBS (smart battery system)-compatible [58] battery ICs, chargers and software drivers. These companies include AMI, Award Software, Hitachi, MCC, Microchip, O2 Micro, Phoenix Technologies, SystemSoft and VLSI Technologies.

Table 2.6. Main battery-management IC producers.

Texas Instruments Battery-management ICs	Philips Semiconductors Charge-control and monitor ICs	Microchip Technology Inc Battery-management ICs
Hitachi America Ltd Power-management µCs	Zilog Inc µC-based charge controller	Integrated Circuit Systems Inc Charge-control ICs
Linear Technology Corp Battery-management ICs	Maxim Integrated Products Charge-control ICs, SMBus-control ICs	National Semiconductor Corp Battery-management ICs
Analog Devices Battery-management ICs	Xicor Battery-management ICs	

Book-keeping Texas Instruments (TI) SoC ICs such as the bq2040, bq26220 [59], bq2060, bq2063 and many others are available on the market. In 2003, TI announced the release of its bq26500 battery fuel gauge – a SoC IC that incorporate an on-board processor to calculate the remaining battery capacity and system run-time (time-to-empty). The device measures the charge and discharge currents using an integrated low-offset voltage-to-frequency converter. A separate ADC on the IC is used to measure battery voltage and temperature. Two count registers accumulate charge and discharge counts, which are generated by sensing the voltage between the two sense resistor pins. Using the measurement inputs, the bq26500 runs a proven book-keeping algorithm (Coulomb counting that is also compensated for self-discharge and discharge rates) to accurately calculate remaining battery capacity and system run-time. The bq26500 compensates remaining battery capacity and run-times for temperature variations. The host system processor simply reads the data set in the bq26500 to retrieve remaining battery capacity, run-time and other critical information that is fundamental to comprehensive battery and power management, including available power, average current, temperature, voltage, time-to-empty and full charge [60]. In the TI ICs, a self-discharge count register counts at a rate of one count every hour at 25°C [61]. The self-discharge count rate doubles approximately every 10°C up to 60°C and is halved every 10°C below 25°C down to 0°C. This value is useful in estimating the battery self-discharge on the basis of capacity and storage temperature conditions.

The Microchip P3 SMBus Smart Battery ICs contain advanced battery control algorithms to determine remaining capacity, run-times and various other data relating to power-management systems [62]. These book-keeping algorithms also rely on a battery cell model that provides performance information of the particular chemical system in use. The cell models are often referred to as 'look-up tables' or LUTs. The look-up tables are stepwise approximations of the performance response curves that can be drawn when looking at discharge performance as a function of for example temperature and discharge rate. There are two look-up tables that define the predictive model for the lithium-ion cell chemistry. One look-up table predicts values of residual capacity (the capacity that cannot be removed from a battery due to discharging efficiency) that are used in remaining time calculations. The second is a predictive model of lithium self-discharge. The self-discharge parameter table predicts self-discharge rates as a function of temperature and total capacity loss. To create look-up tables from cell data enough test points must be generated to create a series of curves that can be broken down into intervals to estimate the curvature with a stepwise approximation. Some LUT 'tables' are illustrated in Fig. 2.16.

The top left graphic in Fig. 2.16 represents the charge efficiency model for NiMH chemistry, the top right represents Li-ion discharge cell models. In the second row of the figure, the discharge performance and self-discharge of NiMH batteries are represented on the left and right, respectively.

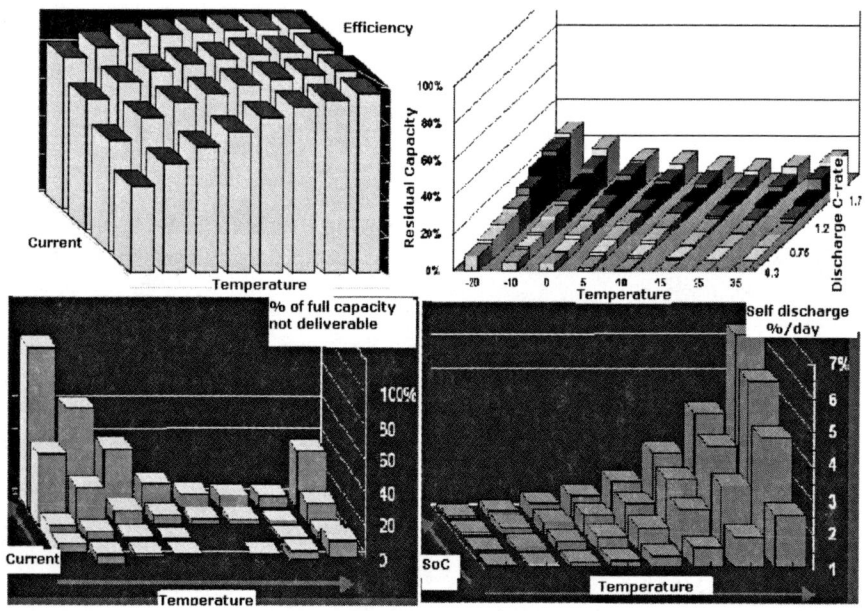

Fig. 2.16. Microchip 3D batteries models [63].

Maxim battery-management devices for Li-ion cells with a Coulomb counter, temperature converter and 15 bytes of user EEPROM, such as the DS2760 high-precision Li-ion battery monitor or 2438 integrated current accumulator (ICA), are available in practice. For the Maxim book-keeping algorithms to function accurately while minimising computational complexity and parametric data storage, certain assumptions are made.

Charge efficiency and pack self-discharge rate are assumed negligible in the case of Li-ion applications and are ignored [48]. The fuel gauging equations work by comparing the ICA value with expected 'empty' and 'full' values for that cell type, which are stored in the IC user EEPROM. These data are generated by characterising the cell type over the application's expected temperature range and current consumption. This information is subsequently stored in a pack-resident memory for the algorithms to later extract and modify. Information should be gathered on several packs so that average or typical values can be stored in every production pack. For best accuracy, the data should be collected on assembled packs containing the production circuit as opposed to individual cells [48].

To collect the data, the cell pack is fully charged at each temperature and fully discharged at each rate and each temperature. All collected data points are arranged as shown in Table 2.7. Since only the difference between points is important, the absolute values of the data do not matter, they have been normalised to the lowest value (standby current empty at 40°C). This reduces the number of data needed to be stored since standby empty 40°C is now always 0.

Table 2.7. Cell characterisation data.

	0°C	10°C	20°C	30°C	40°C
FULL (mAh)	554	561	578	582	588
STANDBY EMPTY (mAh)	65	42	19	11	0
ACTIVE EMPTY (mAh)	124	90	65	50	44

The active empty and standby empty points are defined as the capacity at which the battery reaches the empty voltage (as defined by the user) under the load of the active current and standby current, respectively [64]. The characterisation data are stored in two pages of the IC EEPROM memory. When characterisation of the cell pack is complete, calculating remaining capacity is very simple. The characterisation data are used to find the cell full and empty points on the basis of temperature and discharge rate.

The main drawback of all the discussed ICs is that none of them includes an adaptive method allowing for spread in battery and user behaviour, a large temperature and current range and aging of the cells under all realistic user conditions.

2.7 Conclusions

An overview of the state-of-the-art of State-of-Charge indication of rechargeable batteries, including a historical development of SoC technologies, has been given in this chapter. The general operational mechanism of batteries and the characteristics of the best-known three battery types were given in section 2.2.

The focus in section 2.3 was on the historical development of SoC indication. A general functional architecture of a SoC system was described in section 2.4.

Three basic methods for SoC indication were identified in section 2.5, including that of direct measurement. A particular direct measurement method is the EMF method. The advantage of this method is that the EMF curve does not depend on many parameters. The fact that it does not depend on aging or battery temperature, makes it potentially suitable for State-of-Charge indication. The main drawback of the EMF method is that it does not provide continuous indication of State-of-Charge.

A book-keeping system has been discussed with reference to a mobile phone application. The main problem in the case of book-keeping systems is defining reliable calibration opportunities that occur often enough during a battery's use [1].

Also described above are three of the best-known adaptive methods, which have as inputs measured battery variables such as voltage, impedance and current, and use these variables in order to accurately predict the SoC and the remaining time of use for an application. Using at least one of these adaptive systems should improve a system's ability to cope with aging and temperature effects and spread in battery and user behaviour [1].

The focus in section 2.6 was on commercially available SoC systems. In this section a couple of the actual technologies used by most companies in practice were discussed. The drawback of all the discussed ICs is that none of them includes a real adaptive method allowing for spread in battery and user behaviour, a large temperature and current range and aging of the cells under all realistic user conditions.

2.8 References

[1] H.J. Bergveld, W.S. Kruijt, P.H.L. Notten, Battery Management Systems, Design by Modelling, Philips Research Book Series, **1**, Kluwer Academic Publishers, Boston (2002)
[2] I. Buchmann, Batteries in a Portable World, **2**, (2001)
[3] H. Zijlstra, F.F. Westendorp, Solid State Commun., **7**, 857 (1969)
[4] P.A. Boter, Rechargeable electrochemical cell, US Patent 4,004,943, filed 20 August (1975)
[5] H.J.H. van Deutekom, Rechargeable electrochemical cell, US Patent 4,214,043, filed 4 May (1978)
[6] P.H.L. Notten, J.R.G. van Beek, Nickel–metal hydride batteries: from concept to characteristics, Chem. Ind., **54**, 102–115 (2000)
[7] N. Furukawa, T. Ueda, Sanyo NiMH batteries outperform Li-ion, Nikkei Electronics Asia, 80–83 (1996)
[8] M. Broussely, S. Herreyre, P. Biensan, P. Kasztejna, K. Nechev, R.J. Staniewicz, Aging mechanism in Li-ion cells and calendar life predictions, J. Power Sources, **98**, 13–21 (2001)
[9] R. Spotnitz, Simulation of capacity fade in lithium-ion batteries, J. Power Sources, **113**, 72–80 (2003)
[10] P.M. Gomadam, J.W. Weidner, R.A. Dougal, R.E. White, Mathematical modeling of lithium-ion and nickel battery systems, J. Power Sources, **110**, 267–284 (2002)
[11] B.F.W.H. Heyer, One meter battery tester, US Patent 2,225,051, filed May 16 (1938)
[12] H. Dreer, Curtis wheelchair battery fuel gauge Product Test Report Marketing Services Dept., Curtis Instruments Inc., (1984)
[13] J.J. Kauzlarich, Electric wheelchair fuel gauge tests, Report No UVA-REC-102-86, UVA Rehabilitation Engineering Center (1986)
[14] Curtis Instruments, A Brief History of Innovation and Excellence, http://www.curtisinst.com/index.cfm?fuseaction=c Company.dspHistory, (2005)
[15] E.P. Finger, E.M. Marwell, Battery control system for battery operated vehicles, US Patent 4,012,681, filed 3 January (1975)
[16] S. Lerner, H. Lennon, H.N. Seiger, Development of an alkaline battery state of charge indicator, J. Power Sources, **3**, 135–137 (1970)
[17] R.A. York, Self-testing battery discharge indicator, US Patent 3,932,797, filed 24 December (1974)
[18] S. Kleineberg, M. Kauffmann, Method and circuit arrangement for determining the average current consumption of a battery-operated apparatus, US Patent 2005/0104559 A1, filed 17 February (2003)

[19] S. Hing, Device for estimating the state of charge of a battery, US Patent 6,529,840, filed 12 October (2000)
[20] T.J. Goedken, J.F. Goedken, Method and apparatus for detecting the state of charge of a battery, US Patent 5,185,566, filed 18 November (1991)
[21] R. Brandwein, M.L. Gupta, Nickel–cadmium battery monitor, US Patent 3,940,679, filed 18 June (1974)
[22] H. Laig-Horstebrock, E. Meissner, G. Richter, Method for determining the state of charge and loading capacity of an electrical storage battery, US Patent 6,362,598, filed 26 April (2001)
[23] C.C. Christianson, R.F. Bourke, Battery state of charge gauge, US Patent 3,946,299, filed 11 February (1975)
[24] R.L. Eby, Method and apparatus for determining the capacity of lead acid storage batteries, US Patent 4,180,770, filed 1 March (1978)
[25] S. Rodrigues, N. Munichandraiah, A.K. Shukla, A review of state-of-charge indication of batteries by means of a.c. impedance measurements, J. Power Sources, **87**, 12–20 (1999)
[26] E.J. Dowgiallo Jr., Method for determining battery state of charge by measuring A.C. electrical phase angle change, US Patent 3,984,762, filed 7 March (1975)
[27] E. Zaugg, Process and apparatus for determining the state of charge of a battery, US Patent 4,433,295, filed 8 January (1982)
[28] K. Muramatsu, Battery condition monitor and monitoring method, US Patent 4,678,998, filed 9 December (1985)
[29] E. Peled, H. Yamin, I. Reshef, D. Kelrich, S. Rozen, Method and apparatus for determining the state-of-charge of batteries particularly lithium batteries, US Patent 4,725,784, filed 10 September (1984)
[30] U. Kopmann, Method of and apparatus for monitoring the state of charge of a rechargeable battery, US Patent 4,677,363, filed 30 June (1987)
[31] L. Bowen, R. Zarr, S. Denton, A microcontroller-based intelligent battery system, IEEE AES System Magazine, 16–19 (1994)
[32] E.P. Finger, Quiescent voltage sampling battery state of charge meter, US Patent 4,460,870, filed 23 July (1981)
[33] Y. Tanjo, T. Nakagawa, H. Horie, T. Abe, K. Iwai, M. Kawai, State of charge indicator, US Patent 6,127,806, filed 14 May (1999)
[34] J.H. Aylor, A. Thieme, B.W. Johnson, A battery state-of-charge indicator for electric wheelchairs, IEEE Trans. Indust. Electron., **39**, 398–409 (1992)
[35] T. Kikuoka, H. Yamamoto, N. Sasaki, K. Wakui, K. Murakami, K. Ohnishi, G. Kawamura, H. Noguchi, F. Ukigaya, System for measuring state of charge of storage battery, US Patent 4,377,787, filed 8 August (1980)
[36] G.R. Seyfang, Battery state of charge indicator, US Patent 4,949,046, filed 21 June (1988)
[37] M.W. Verbrugge, E.D. Tate Jr, S.D. Sarbacker, B.J. Koch, Quasi-adaptive method for determining a battery's state of charge, US Patent 6,359,419, filed 27 December (2000)
[38] G. Richter, E. Meissner, Method for determining the state of charge of storage batteries, US Patent 6,388,450, filed 15 December (2000)
[39] D. Stolitzka, W.S. Dawson, When is it intelligent to use a smart battery?, No 94 TH0617-1 IEEE, (1994)

[40] O. Gerard, J.N. Patillon, F. d'Alche-Buc, Neural network adaptive modelling of battery discharge behaviour, Lect. Notes Comput. Sci., **1327**, 1095–1100 (1997)
[41] A.J. Salkind, C. Fennie, P. Singh, T. Atwater, D.E. Reisner, Determination of state-of-charge and state-of-health of batteries by fuzzy logic methodology, J. Power Sources, **80**, 293–300 (1999)
[42] J. Garche, A. Jossen, Battery management systems (BMS) for increasing battery life time, Telecommunications Energy Special, **3**, 81–84 (2000)
[43] H.J. Bergveld, H. Feil, J.R.G.C.M. Van Beek, Method of predicting the state of charge as well as the use time left of a rechargeable battery, US Patent 6,515,453, filed 30 November (2000)
[44] F.A.C.M. Schoofs, W.S. Kruijt, R.E.F. Einerhand, S.A.C. Hanneman, H.J. Bergveld, Method of and device for determining the charge condition of a battery, US Patent 6,420,851, filed 29 March (2000)
[45] P.P.L. Regtien, F. van der Heijden, M.J. Korsten, W. Olthuis, Measurement Science for Engineers, Kogan Page Science Publisher, London (2004)
[46] D. Linden, Handbook of batteries, 2nd edition, McGraw-Hill, New York, (1995)
[47] Dallas Semiconductors, Inaccuracies of estimating remaining cell capacity with voltage measurements alone, Application Note 121, (2000)
[48] Dallas Semiconductors, Lithium-ion cell fuel gauging with Dallas semiconductor devices, Application Note 131, (2000)
[49] G.A. Nazri, G. Pistola, Science and Technology of Lithium Batteries (2002)
[50] P.P.L. Regtien, Instrumentation Electronics, Prentice-Hall, New York, (1992)
[51] F. Huet, A review of impedance measurements for determination of the state-of-charge or state-of-health of secondary batteries, J. Power Sources, **70**, 59–69 (1998)
[52] Philips Internal Communications, (2005)
[53] G. Plett, Extended Kalman filtering for battery management systems of LiPB-based HEV battery packs: Part 1. Background, J. Power Sources, **134**, 252–261 (2004)
[54] G. Plett, Extended Kalman filtering for battery management systems of LiPB-based HEV battery packs: Part 2. Modeling and identification, J. Power Sources, **134**, 262–276 (2004)
[55] G. Plett, Extended Kalman filtering for battery management systems of LiPB-based HEV battery packs: Part 3. State and parameter estimation, J. Power Sources, **134**, 277–292 (2004)
[56] S. Grewal, D.A. Grant, A novel technique for modelling the state of charge of lithium ion batteries using artificial neural networks, Proc. Int. Telecommunications Energy Conf. (IEEE), **484**, 14–18 (2001)
[57] S. Piller, M. Perrin, A. Jossen, Methods for state-of-charge determination and their applications, J. Power Sources, **96**, 113–120 (2001)
[58] D. Friel, M. Mattera, SBS smart battery interface guidelines, Portable and Wireless (1998)
[59] Texas Instruments, High-performance battery monitor IC with Coulomb counter, voltage and temperature measurements, Doc. I.D. SLUS521A (2002)
[60] Texas Instruments, Single-cell Li-ion and Li-Pol battery gas gauge IC for handheld applications (bq junior family), Doc. I.D. SLUS567A (2003)
[61] Texas Instruments, High-performance battery monitor with coulomb counting and flash memory, Doc. I.D. SLUS509 (2001)

[62] Power Smart, P3 cell model explanations, Application Note 450799 (1999)
[63] L. Conklin, Smart battery data accuracy, Intel Developer Forum (1999)
[64] Dallas Semiconductors, Characterizing a Li+ cell for use with a fuel cell, Application Note 2412 (2003)

Chapter 3
A State-of-Charge indication algorithm

As discussed in chapter 2, many advances have been made in State-of-Charge (SoC) indication in recent years, both through continued improvement of the SoC algorithms and through the development of more accurate hardware systems. Nevertheless, there is still no "ideal" SoC system that gives accurate indications under all realistic user conditions. The "ideal" SoC system is obviously one that is not expensive, can handle all battery chemistries, can operate over a wide range of load currents and can deal with the aging effect. Leading semiconductor companies (e.g. Philips [1]–[3], NXP Research, Texas Instruments [4]–[6], Microchip [7], [8] Maxim [9], [10], etc.) are paying more and more attention to accurate State-of-Charge indication in attempts to find that ideal system.

A SoC algorithm that combines some form of adaptivity with direct measurement and book-keeping systems was developed and implemented by Bergveld et al. in 2000 [1]–[3]. By implementing the mathematical models described in [1], this algorithm was found to be the most sophisticated and accurate [11], [12]. This chapter will give a complete description of this algorithm, which serves as the starting point of this book. This chapter is organised as follows. An introduction to the algorithm is given in section 3.1. Section 3.2 describes the models and states of the SoC indication system. The main aspects of the algorithm are given in section 3.3. The focus in section 3.4 is on accuracy problems. Section 3.5 presents concluding remarks.

3.1 An introduction to the algorithm

The SoC indication algorithm presented by Bergveld *et al.* in [1]–[3] aims to eliminate the main drawbacks and combine the advantages of the direct measurement and book-keeping methods described in Chapter 2. The basis of the SoC algorithm is Electro-Motive Force (EMF) measurement during equilibrium and current measurement and integration during charge and discharge. During discharge, in addition to simple Coulomb counting, the effect of the overpotential is also considered [1]. A method has also been developed for updating the value of the maximum capacity for coping with capacity loss due to the aging effect. The algorithm will be described below for a Panasonic CGR17500 Li-ion battery, but the basis of the algorithm holds for other types of Li batteries, too. The rated capacity of this battery is 720 mAh.

3.2 Battery measurements and modelling for the State-of-Charge indication algorithm

The battery model applied in the developed SoC indication algorithm describes the battery EMF and overpotential behaviour, neither of which can be measured directly. The EMF and overpotential curves have been measured with an accurate battery tester and implemented in the Battery Management System (BMS) using mathematical-function approximations [1], [13]. Both the measurement and the implementation method contribute to the final accuracy of the SoC indication. The EMF and overpotential measurement and modelling methods used in [1] will be described further on in this section.

3.2.1 EMF measurement and modelling

Two measurement methods for EMF determination are considered in [1]: *voltage relaxation* and *linear interpolation*. A comparison with the EMF curves obtained in voltage relaxation and interpolation is shown in Fig. 3.1 [1]. The voltage relaxation measurements were performed by charging and discharging a battery in small 15 mAh increments at a rate of 0.1 C. Each charge and discharge was followed by a rest period of 30 minutes, after which the voltage was sampled. This voltage was assumed to be equal to the EMF. The battery was charged and discharged in 48 steps. The battery was charged with a current of 0.1 C-rate up to 4.1 V in the interpolation method. After a rest period of 30 minutes, the battery was discharged at a rate of 0.1 C to 3 V. In order to average the results of the charging and discharging cycles, each nth discharge curve was averaged with each $(n+1)^{st}$ charge curve.

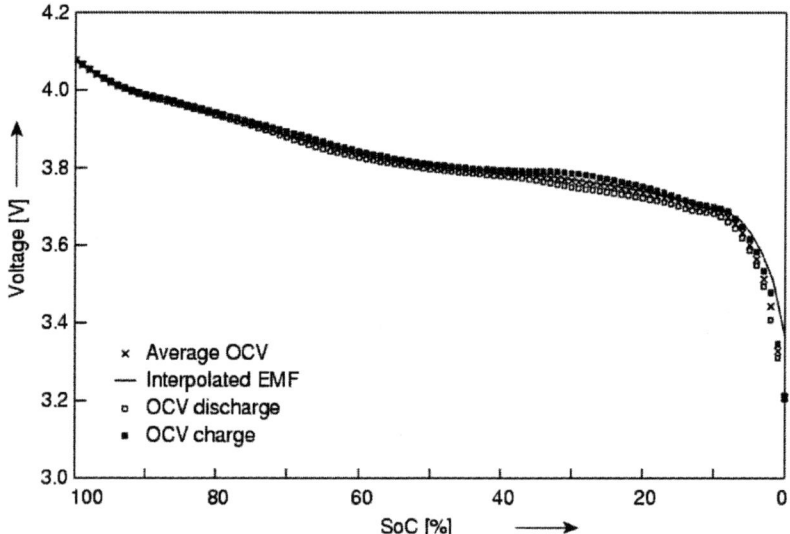

Fig. 3.1. Comparison of the EMF curve obtained in voltage relaxation and interpolation measurements. The EMF values obtained in the voltage relaxation measurements are represented as squares; measurement points were obtained after charging (■) and discharging (□). The curve (x) represents the mathematical average of the (■) and (□) curves. The solid line represents the EMF curve obtained in interpolation. The x-axis shows SoC [%] normalised to maximum capacity; all measurements were performed at 25°C [1].

Fig. 3.1 shows that the two curves obtained in voltage relaxation after charge and discharge steps are not identical. This is due to the hysteresis effect [1], [14]–[17]. The greatest hysteresis occurs at a capacity of roughly 30% and amounts to approximately 40 mV (around 12% SoC when only the EMF curve obtained after discharge steps is taken into account in an SoC indication system). More information on the EMF hysteresis will be given in chapter 4 of this book. The interpolated EMF and the mathematical average of the voltage relaxation curves are practically identical. Only a small deviation occurs when the battery is almost empty, which is partly due to longer relaxation times needed during discharging. In this case rest times of 30 minutes were not enough [1].

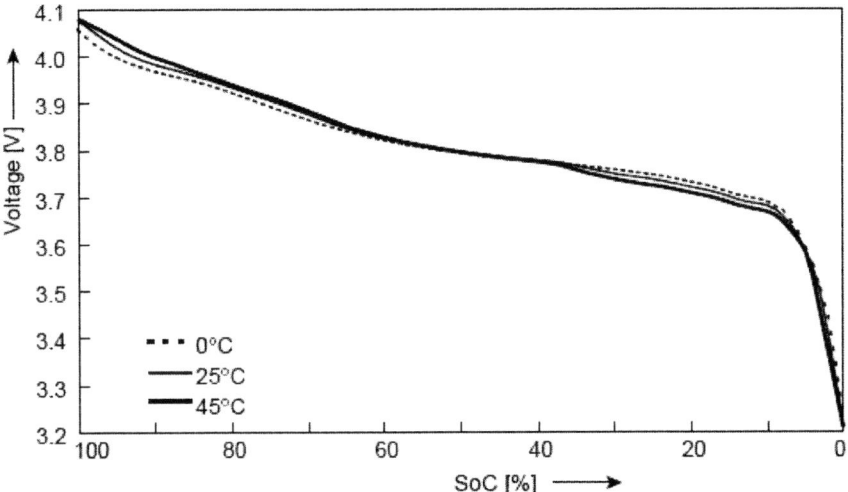

Fig. 3.2. Measured EMF curves obtained in voltage relaxation after discharge steps at different temperatures: 0°C, 25°C and 45°C. The x-axis shows SoC [%] normalised to maximum capacity [1].

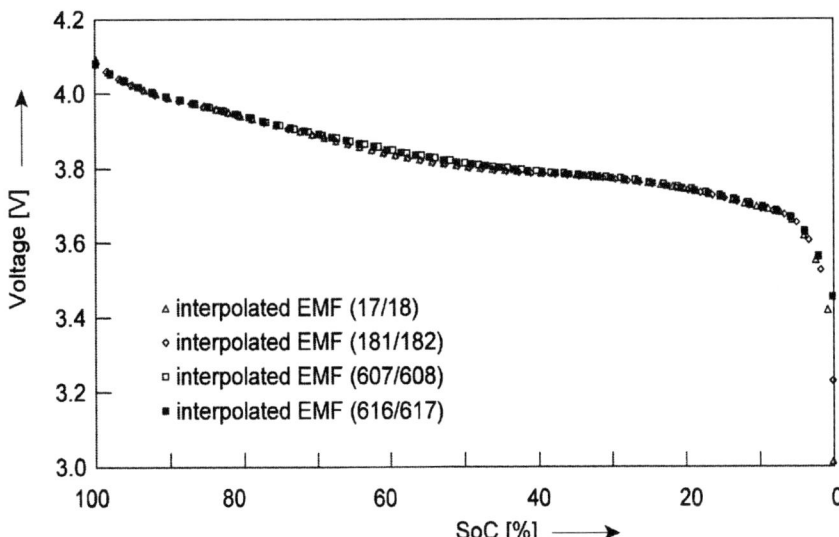

Fig. 3.3. Measured interpolated EMF curves obtained for different charge/discharge cycles at 25°C. Cycle 17/18 (Δ), cycle 181/182 (◊), cycle 607/608 (□), cycle 616/617 (■). The x-axis shows SoC [%] normalised to maximum capacity [1].

Fig. 3.2 shows the EMF curves obtained in voltage relaxation after discharge steps at different temperatures [1]. It illustrates that the temperature dependence of the EMF is almost zero at 50% SoC. At SoC values higher than 50%, the EMF decreases at low temperatures. At values lower than 50% SoC, the EMF increases at low temperatures. The maximum error in the SoC in the temperature range from 0°C to 45°C occurs around 30% SoC and amounts to 8% when only the EMF curve at 25°C is taken into account in an SoC indication system.

Fig. 3.3 shows EMF curves interpolated for different charge/discharge cycles. The EMF curves are practically the same for all cycle numbers when plotted on the normalised capacity axis. The maximum difference is only 10 mV. The EMF curves plotted with an absolute capacity axis will differ, because the absolute battery capacity decreases when a battery ages. The absolute capacity values that were found for cycles 14, 289 and 615 were 733 mAh, 707 mAh and 660 mAh, respectively. It can be concluded from Fig. 3.3 that the EMF curve depends on the aging effect to only to a limited extent, which makes it potentially suitable for SoC indication. The EMF method can also be used to calibrate the SoC system because in charge and discharge states the calculated SoC will eventually drift away from the real value due to *e.g.* measurement inaccuracy in the current and the integration in time of this inaccuracy.

The EMF curves were further simulated using a model similar to the one described in section 4.4 of this book. Temperature dependence was taken into account only in the term RT/F [1].

3.2.2 Overpotential measurement and modelling

The second model presented here is the overpotential model. A battery's voltage during the discharge state is lower than the EMF due to overpotential. As overpotential represents the difference between EMF and the discharge/charge battery voltage, an EMF should first be determined. The method presented in [1] considers a battery discharge curve at 0.1 C-rate as a reference for the overpotential calculation. The difference between the reference curve and different discharge curves at current rates of 0.2 C-rate, 0.4 C-rate, 0.8 C-rate and 1 C-rate, respectively, has been considered for the overpotential interpretation [1].

The overpotential prediction also yields remaining run-time prediction. When current is drawn from a battery during discharging overpotentials occur. A battery will appear empty to a user even if a certain amount of capacity is still present inside the battery, because the battery voltage drops below the End-of-Discharge voltage (V_{EoD}) defined in a portable device (*e.g.* 3 V in the case of a Li-ion battery). This is illustrated in Fig. 3.4, which shows the remaining run-time (t_r) plotted along the horizontal axis to explain this effect.

A State-of-Charge indication algorithm

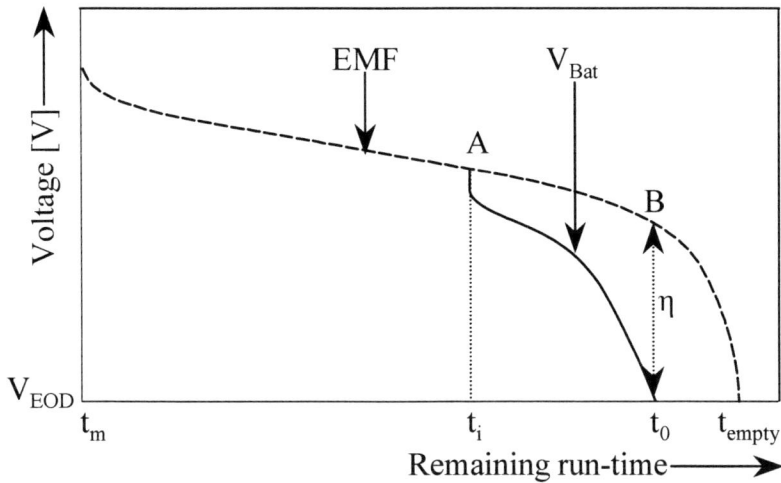

Fig. 3.4. Schematic representation of EMF (dashed) and discharge voltage (solid) curves leading to an empty battery at $t_r = t_{empty}$. The battery appears empty at t_0. The x-axis shows the remaining run-time.

As can be seen in Fig. 3.4, discharging starts at point A and the battery voltage drops at an overpotential η that is a function of the discharge current I and temperature T. At this moment a remaining run-time $t_r = t_i$ is calculated on the basis of the following equation:

$$t_r(I,T) = \frac{Q_A - Q_B}{I} \qquad (3.1)$$

where Q_A [C] is the battery capacity at the beginning of discharging at point A and Q_B [C] represents the battery capacity at point B calculated on the basis of:

$$Q_B = \frac{SoC(EMF_B)}{100} * Q_{max} \qquad (3.2)$$

where $SoC(EMF_B)$ [%] represents the SoC calculated on the basis of the estimated EMF at point B and Q_{max} represents the battery's maximum capacity. The estimated EMF_B is a sum of the End-of-Discharge voltage and the predicted overpotential η.

At point B, $t_r = t_0$ and the remaining capacity is zero under the present I and T conditions. The battery will be completely empty (point $t_r = t_{empty}$ in Fig. 3.4) when the battery voltage reaches the End-of-Discharge voltage and the overpotential equals zero. Hence, a distinction should be made between the charge available in a battery (*i.e.* SoC) and the charge that can be withdrawn from the battery under certain conditions, expressed in remaining run-time. It can be concluded that prediction of the overpotential at point B will also yield a more accurate remaining run-time prediction.

Four types of overpotential are identified in [1]: Li^+ transport overpotential in the electrolyte, total kinetic or reaction overpotential, Li^+ diffusion overpotential in both the $LiCoO_2$ electrode and the LiC_6 electrode. On the basis of this identification an overpotential model that considers the ohmic, kinetic ($\eta_{\Omega k}$), diffusion (η_d) and increase in the overpotential when the battery becomes empty (η_q) has been formulated as follows [1]:

$$\eta(Q_{in},T,I,t) = \eta_{\Omega k}(T,I,t) + \eta_d(T,I,t) + \eta_q(Q_{in},T,I,t) = $$
$$I*(R_{\Omega k}(T) + R_d(T)*(1-e^{\frac{-t}{\tau_d(T)}}) + \frac{U_q(I)*E_q(T)}{Q_{in}(t)}*(1-e^{\frac{-t}{\tau_q(T)}})) \quad (3.3)$$

where η is the total overpotential (in [V]), I denotes the applied discharge current (in [A]), $R_{\Omega k}(T)$ is the temperature-dependent "ohmic" and "kinetic" resistance (in [Ω]), $R_d(T)$ is the temperature-dependent "diffusion" resistance (in [Ω]) and $\tau_d(T)$ denotes the temperature-dependent "diffusion" time constant (in [s]). The dimensionless function $U_q(I)$ is an inverse step function, with $U_q(I) = 1$ for discharging ($I \leq 0$) and $U_q(I) = 0$ for charging ($I > 0$). The variable $Q_{in}(t)$ is the charge present inside the battery at time t, which means that Q_{in} expresses SoC in [C]. Finally, $\tau_q(T)$ is a temperature-dependent time constant associated with the increase in overpotential in an almost empty battery (in [s]) [1]. Simulation results of the mathematical overpotential implementation showed good agreement with the measured overpotential data [1].

3.3 States of the State-of-Charge algorithm

The algorithm proposed in [1]–[3] operates in five different states: *initial state, equilibrium state, transitional state, discharge state* and *charge state*. The state diagram illustrating the basic structure of the algorithm is shown in Fig. 3.5.

When a battery is first connected to the SoC system, the algorithm will start up in the *initial state*. In this state the initial SoC is determined on the basis of voltage and temperature measurements and the stored SoC-EMF relationship. This initial SoC is shown to the user, as it is assumed to be unacceptable for users to have to wait for more than a few seconds before being able to check the available capacity after the system has been switched on. Depending on whether the battery is charged, discharged or in equilibrium, the algorithm will then shift to the appropriate state.

In the *equilibrium state* hardly any current is drawn from the battery. This situation will for example occur when a mobile phone is in standby mode. The current will in this case be only a few mA, which is lower than a small current I_{lim} defined in the system (*e.g.* 2 mA in a mobile phone application). At this very low current value, the battery voltage will be very close to the EMF value, providing that the voltage is stable. So, stable voltage is necessary to allow the algorithm to change to this state. In this state the SoC is determined on the basis of voltage measurements and the stored SoC-EMF relationship.

A State-of-Charge indication algorithm

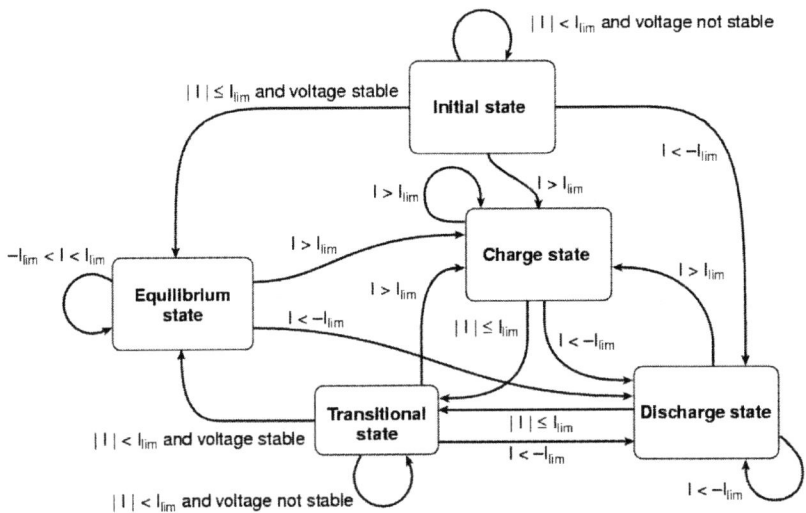

Fig. 3.5. State diagram of the SoC algorithm [1].

The *transitional state* is used when the algorithm changes from either the *charge* or the *discharge state* to the *equilibrium state*. In this state it is determined whether the battery voltage is stable and the algorithm is allowed to enter the *equilibrium state*. This is achieved by calculating the derivative of the battery voltage over time dV/dt and comparing it with a threshold value dV/dt_{lim} stored in the SoC system to check the voltage stable condition [1]. The dV/dt_{lim} threshold should be chosen to be small enough to ensure that the battery voltage is constant in time and that the battery reaches the *equilibrium state*.

In the *charge state*, a charger is connected to the battery and a positive current larger than I_{lim} flows into the battery. The SoC is determined by Coulomb counting. At the end of the *charge state*, the system passes through the *transitional state* to the *equilibrium state*.

In the *discharge state*, the battery is discharged and a negative current larger in module than I_{lim} flows out of the battery. In addition to simple Coulomb counting the effect of the overpotential is also considered. As has been shown in section 3.2 the prediction of the overpotential also yields a remaining run-time prediction. At the end of the *discharge state*, the system passes through the *transitional state* to the *equilibrium state*.

In practice, any battery will lose capacity during cycling. A simple method for updating the maximum capacity Cap_{max} to take capacity loss into account is described in [1] (see also Fig. 3.6).

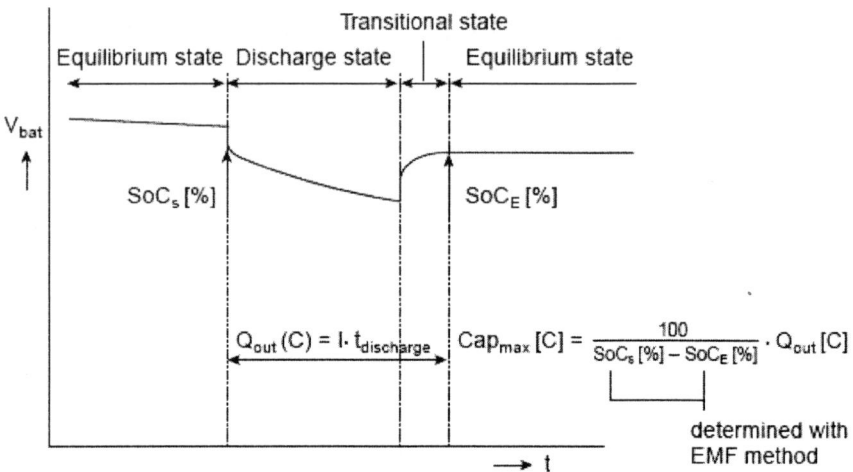

Fig. 3.6. Simple method for updating Q_{max} to take capacity loss into account [1].

As shown in Fig. 3.6, for the mechanism for updating the maximum capacity it is necessary for the system to run through a sequence of states: *equilibrium state, discharge state, transitional state* and *equilibrium state*. The new value of Cap_{max} is simply calculated by relating the charge Q_{out} drawn from the battery mainly in the discharge state and to a much lesser extent in the transitional state to the difference in SoC (SoC_S-SoC_E) before and after discharging. The Q_{out} is measured with the Coulomb counting method. The SoC_S and SoC_E values are inferred from a stored EMF-SoC relationship. The method can be made more complex by enforcing a minimum value of Q_{out} for the update to be valid. Moreover, only small changes in Cap_{max} should be allowed. Finally, the update mechanism should be allowed to occur only under "standard" conditions, for example a discharge rate that is not higher than 1 C-rate and a temperature that is within the range from 10°C to 40°C. A similar updating mechanism can be implemented during charging [1].

In summary, in which state the algorithm is operating will depend on the value and sign of the current flowing into or out of the battery and on whether the battery voltage is stable. The main aspects of the described SoC algorithm will be discussed in the next section of this chapter.

3.4 Main issues of the algorithm

An accurate battery model is essential for accurate SoC indication. As has been shown in the previous sections, the model applied in the SoC indication algorithm described in [1]–[3] combines maximum capacity adaptation with a battery's EMF and overpotential behaviour. The EMF and overpotential curves are measured indirectly by a battery tester [1]. The main aspects of the measurement, modelling and implementation methods applied in [1] for EMF, overpotential and maximum capacity adaptation will be further discussed in this section. This will lead to identification and better understanding of the SoC inaccuracy sources, which will be useful for improving SoC system accuracy.

3.4.1 EMF measurement, modelling and implementation

Two measurement methods have been considered for EMF determination: *voltage relaxation* and *linear interpolation*. A rest period of 30 minutes has been considered to obtain an EMF value by means of *voltage relaxation* [1].

Fig. 3.7 illustrates what happens to the battery voltage after application of a discharge step. As can be seen, during the transition process, a battery's OCV (Open-Circuit Voltage) doesn't instantaneously coincide with its EMF, but relaxes to it. Especially at low temperatures this may take a long time; see Fig. 3.7. The value of the OCV changes from 3.680 V immediately after the current interruption and to about 3.729 V after 480 minutes. It follows from Fig. 3.7 that the OCV is constant after about 400 minutes. The voltage after 30 minutes differ approximately 4 mV from the voltage after 480 minutes. This means that if it is assumed in this example that the equilibrium state is reached after 30 minutes, the inaccuracy in the SoC indication will be about 1%, as can be inferred from the sensitive part of the EMF curve. This situation will be even worse in the case of aged batteries at low SoC and temperature values [12]. It can be concluded that a 30 minutes' rest period is not enough for accurate EMF curve determination.

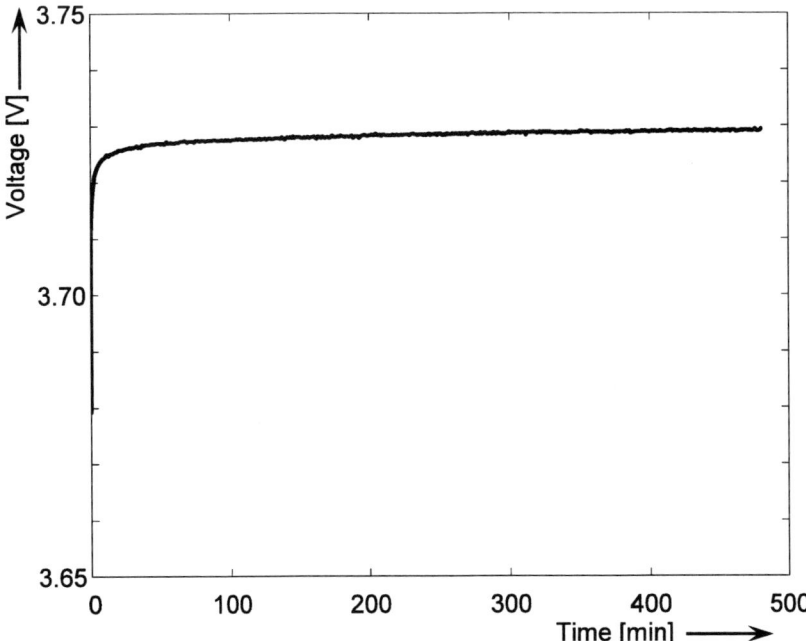

Fig. 3.7. Voltage relaxation after a discharge current step of 0.1°C-rate at 18% SoC and 5°C. The x-axis represents the time of the transitional state in minutes.

The *linear interpolation* method considers the average battery voltage of the charge and discharge cycles at the same SoC and temperature in order to obtain the EMF curve. The battery voltage is averaged to minimise the overpotential influence and the hysteresis effect in the EMF determination method [14]–[17]. However, as shown in [18], the EMF hysteresis causes a maximum difference of 30 mV (about 8% SoC) at around 30% SoC between the EMF measured by means of *voltage relaxation* after charge steps and the EMF measured by means of *voltage relaxation* after discharge steps. When the EMF obtained by means of *linear interpolation* is considered in this example the inaccuracy of the SoC indication will be at least 4%. It can be concluded that the *linear interpolation* method is not an accurate solution for EMF determination.

It is concluded in [1] that the EMF depends on the aging effect to only a limited extent when plotted on a relative SoC scale. For an analysis of this conclusion the maximum battery capacity determination method applied in [1] must be explained. The maximum battery capacity has been defined as the capacity obtained from a battery during discharging to 3 V at a C-rate of 0.1 after charging to 4.1 V at a rate of 0.1°C. The maximum capacity value is used to plot the EMF curves on a relative axis in [1]. For an accurate calculation of a battery's maximum capacity, two SoC states with corresponding EMF voltages should be identically defined for a fresh and an aged battery. The two defined battery SoC states are an empty and a full battery SoC. An empty battery state, or 0% SoC, may be assumed for example when the equilibrium voltage has a value of 3 V. A full battery state, or 100% SoC, may be assumed for example when the battery voltage has a value of 4.1 V. However, as mentioned in [1], the battery overpotential increases with aging. This means that when the same voltage limits and current C-rates are used during charging/discharging a fresh and an aged battery the calculated maximum battery capacity value will be influenced by the increase in the battery's overpotential. In conclusion, the maximum capacity value measured in [1] was also influenced by the increase in the overpotential. When the overpotential effect is eliminated from the maximum capacity calculation the "real" maximum battery capacity will have a higher value than the value measured in [1]. In this situation the compared EMF-SoC curves, also illustrated in Fig. 3.3 of this chapter, may represent a smaller difference in maximum battery capacity. This conclusion can also explain the similarity between the EMF curves obtained for a fresh and an "aged" battery. More measurements are needed to check the influence of aging on the EMF-SoC relationship. This will be done in chapter 6 of this book.

The temperature influence is simply considered in the factor RT/F in the EMF model presented in [1]. However, each parameter of the EMF model may be temperature-dependent. In conclusion, in order to obtain more accurate SoC indication, an EMF model in which the influence of temperature on the EMF model parameters is included should be further considered. Another factor that should be considered is the EMF model dependency on the battery chemistry. The presented EMF model was developed for a Panasonic CGR17500 Li-ion battery. The positive electrode of this type of battery consists of $LiCoO_2$ and the negative electrode is made of LiC_6. If *e.g.* another lithium metal oxide (*e.g.* $LiNiO_2$ or $LiMn_2O_4$) were to be used to store the lithium ions, the EMF-SoC relationship could be different. For this reason an adaptive method needs to be used that also takes into consideration different battery chemistries for the EMF model description. In conclusion, an adaptive solution will also allow extension of the EMF model applicability to other types of battery chemistries. This will be discussed in chapter 8 of this book.

During implementation of the EMF detection method, false entries into the equilibrium state were detected [1]. This influenced the EMF estimation and the system's calibration accuracy [1]. For a proper understanding of this problem the EMF detection method applied in [1] must be explained. In this method the derivative of the battery voltage over time dV/dt is calculated and compared with a threshold value dV/dt_{lim} stored in the SoC system in order to determine the battery

A State-of-Charge indication algorithm   57

voltage stable condition. In most practical applications a battery will not reach a completely relaxed state because a certain amount of current will always be flowing into or out of the battery. A good example is the standby current in a mobile phone application. As shown in [19] a dV/dt_{lim} threshold that is dependent on the SoC and temperature should be chosen for accurate equilibrium state detection. However, as has been further proven, even when longer t times are chosen, the dV/dt may have a value close to zero while the voltage is still relaxing. In conclusion, it is very difficult to distinguish between a stable and an unstable battery voltage condition by considering only dV/dt measurements. Also, when the same threshold value dV/dt_{lim} is considered for equilibrium state detection in the case of all SoC values the risk of false detections seems to be quite high. More specifications for equilibrium state detection will be given in the next chapter of this book.

3.4.2 Overpotential measurement, modelling and implementation

The overpotential measurement method applied in [1] regards the overpotential as a difference between a reference battery discharge curve obtained at 0.1 C-rate and the battery discharge curves obtained at 0.2, 0.4, 0.8 and 1 C-rate. In conclusion, the measurement method assumes that the battery overpotential (η) is linear as a function of the discharge current rates. A clear definition of battery overpotential is needed to analyse this assumption. By definition, a battery's overpotential represents the difference between the battery's EMF and the charge/discharge battery voltage.

Fig. 3.8 illustrates the battery overpotentials measured at 0.25 and 0.5 C-rate and the calculated battery overpotential at 0.25 C-rate. The measured overpotentials were obtained as a difference between a measured EMF and measured discharge curves at 0.25 and at 0.5 C-rate discharge current rates. The measurements were carried out at 25°C. The calculated battery overpotential at 0.25 C-rate was obtained as a difference at the same SoC, between the overpotential measured at 0.5 C-rate and the overpotential measured at 0.25 C-rate. As can be seen in Fig. 3.8, the calculated overpotential at 0.25 C-rate differs from the measured overpotential at 0.25 C-rate. It can be concluded from Fig. 3.8 that the battery overpotential is not linear as a function of the discharge current rates.

The value of the overpotential depends on the discharge current, the SoC, age and temperature. Especially in the case of old cells, the remaining charge cannot be withdrawn from a battery at low temperatures and low SoC values due to a high overpotential, because otherwise the battery voltage would drop below the End-of-Discharge voltage [1]. In conclusion, the overpotential model parameters should be C-rate-temperature-and age-dependent. The model developed in [1] does not include temperature and age dependence in the overpotential description. Another important factor is the overpotential model dependency on the battery chemistry. Like the EMF model, the overpotential model was also developed for a CGR17500 Li-ion battery. The model parameters were obtained by means of curve fitting using measurements obtained for batteries of this type. An important study will be a comparison between the overpotential behaviour of a CGR17500 Li-ion type of battery with that of a Li battery with a different chemistry.

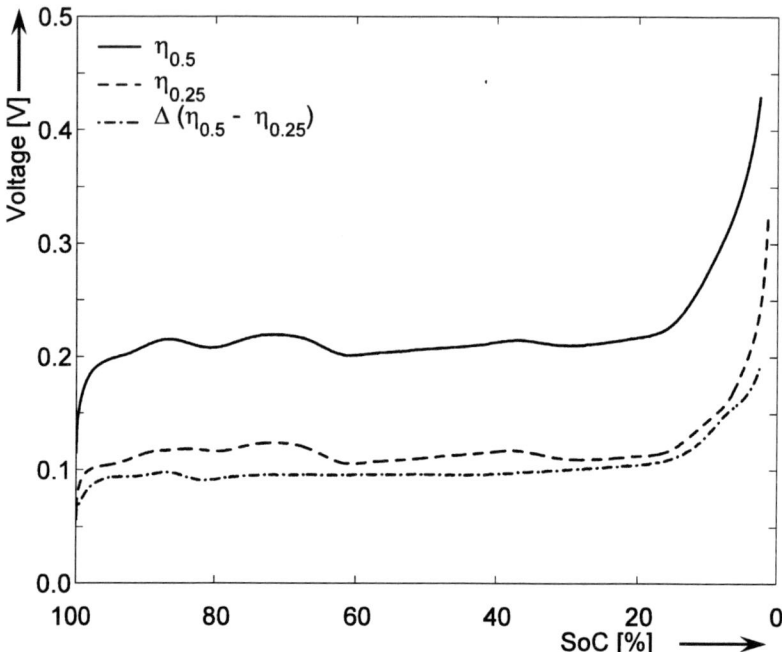

Fig. 3.8. Measured battery overpotentials at 0.25 ($\eta_{0.25}$) and 0.5 C-rate ($\eta_{0.5}$) and calculated battery overpotential at 0.25 C-rate Δ ($\eta_{0.5} - \eta_{0.25}$). The x-axis shows SoC [%], normalised to maximum capacity.

3.4.3 Adaptive systems

To obtain an accurate SoC indication system, the unpredictability of the behaviour of both batteries and users should be considered. This unpredictability necessitates the use of an adaptive system. As already mentioned above, the EMF and the overpotential models should include an adaptive method in order to deal with the aging effect and different battery chemistries. More accurate measurements of the EMF dependence on the aging effect are needed. Such measurements are important for batteries that have a considerable capacity loss. If a battery's EMF will change due to the aging effect then the method considered in [1] for updating the maximum capacity will probably not yield accurate results. As already mentioned above, the overpotential model should also include an adaptive method to adapt the system parameters to the capacity loss. In conclusion, an important improvement in SoC and remaining run-time indication accuracy can be achieved by implementing new adaptive solutions.

Finally, more tests of the SoC system under different conditions, *e.g.* at different charge/discharge currents and temperatures with aged and different batteries chemistries, will be necessary to check the SoC algorithm accuracy.

3.5 General remarks on the accuracy of SoC indication systems

SoC indication, for example in the form of a remaining run-time indication, must be accurate to within less than one minute under all practical conditions, including aging of the battery, for the system to be perceived as reliable by its users. Before considering improvements and recent developments in the fields of SoC and remaining run-time indication we will first further consider the accuracy of SoC systems in the remainder of this chapter.

A system's accuracy is based on a qualification of the expected closeness of the result of a measurement to the actual value [20]. The generally accepted quantitative measure relating to accuracy is uncertainty. The uncertainty of a measurement is a parameter that characterises the range of the values within which a measurement's actual value lies. It expresses our lack of knowledge of the measurand (the particular quantity to be measured) [20].

Several types of errors are covered in SoC indication systems. Typical errors in SoC indication systems are errors in the voltage and temperature measurements, errors in the current measurement and integration in time of these errors. A measurement's error is composed of two components, which are together responsible for the combined uncertainty of the measurement. These components are the random and the systematic errors [1], [20]. The random errors in an SoC indication system are caused by spread in battery behaviour and by measurement inaccuracy. These errors can lead to a probability distribution around the estimated SoC and remaining capacity values. One way of minimising the effects of such errors is by averaging the results of repeated measurements. The systematic error is the error that persists even after averaging of an infinite number of repeated measurements. An example of a systematic error in an SoC indication system is an incorrect or incomplete inclusion of battery behaviour in a system [1]. Systematic errors may lead to a situation in which a portable device stops operating while the estimated remaining capacity is still larger than zero. Once a systematic error is known, it can be quantified and corrected, for example by calibrating the voltage, current and temperature measurements.

To assess the combined uncertainty in an SoC indication system, all possible error sources must first be identified. This can be done for example with the aid of statistical inference techniques (*e.g.* curve fitting) or by means of a pool of information on the variability of the measurement results (*e.g.* the specifications provided by a device's manufacturer) [20]. A next step is to check how these errors propagate through the measurement chain and how they affect the final SoC indication inaccuracy. This is systematically done by means of the error budget, *e.g.* a table that catalogues all contributions to the final error. For further information on the subject of accuracy the reader is referred to [20].

The general remarks concerning uncertainty and error identification presented in this section will be of particular importance in Chapter 7, with respect to a newly developed SoC algorithm.

3.6 Conclusions

In this chapter a State-of-Charge indication system has been presented that calculates the SoC in percentages and also indicates the remaining run-time of a

portable application powered by a Li-ion battery [1]. This algorithm serves as the starting point of this book.
- The main problem involved in EMFs obtained by means of *voltage relaxation* is that a longer rest period is necessary for an accurate EMF curve determination. The main problem involved in EMFs obtained by means of *linear interpolation* is the hysteresis effect. A simplified EMF model has been considered for the EMF-SoC relationship implementation. To enhance the accuracy of the SoC indication system, the influence of battery temperature must be included in the EMF model. And a new EMF detection method should be developed. New measurements, modelling and detection methods for battery EMFs will be presented in chapter 4.
- As discussed in section 3.3, during current flow, Coulomb counting and overpotential prediction are considered in the SoC system [1]. The prediction of the overpotential also yields remaining run-time prediction. As concluded in section 3.4, the difference between two discharge curves does not yield an accurate overpotential determination. Another conclusion relates to the developed overpotential model. This model should further include the influence of temperature and C-rate to improve the accuracy of the SoC and the remaining run-time indication. New measurements and modelling methods for battery overpotentials will be presented in chapter 5.
- The effect of a battery's aging on the accuracy of the EMF and overpotential models needs to be studied further, also with a view to validation of the maximum capacity adaptation method presented in [1]. In this method it is assumed that the EMF curve depends on the aging effect to only a limited extent. Adding more adaptive solutions to the SoC indication system will improve the system's ability to cope with the aging effect and to sustain the accuracy of the SoC and the remaining run-time indication. New measurements of the battery aging effect and new adaptive systems will be presented in chapter 6.
- Finally, more tests under different conditions (*e.g.* different temperatures and C-rates and using batteries of different ages) will have to be carried out to investigate the accuracy of the SoC indication system in greater detail. This will be done in chapters 7 and 8.

3.7 References

[1] H.J. Bergveld, W.S. Kruijt, P.H.L. Notten, Battery Management Systems, Design by Modelling, Philips Research Book Series, **1**, Kluwer Academic Publishers, Boston (2002)

[2] H.J. Bergveld, H. Feil, J.R.G.C.M. Van Beek, Method of predicting the state of charge as well as the use time left of a rechargeable battery, US Patent 6,515,453, filed 30 November (2000)

[3] F.A.C.M. Schoofs, W.S. Kruijt, R.E.F. Einerhand, S.A.C. Hanneman, H.J. Bergveld, Method of and device for determining the charge condition of a battery, US Patent 6,420,851, filed 29 March (2000)

[4] Texas Instruments, High-performance battery monitor IC with Coulomb counter, voltage, and temperature measurements, Doc. I.D. SLUS521A (2002)

[5] Texas Instruments, Single-cell Li-ion and Li-Pol battery gas gauge IC for handheld applications (bq junior family), Doc. I.D. SLUS567A (2003)

[6] Texas Instruments, High-performance battery monitor with coulomb counting and flash memory, Doc. I.D. SLUS509 (2001)
[7] Power Smart, P3 cell model explanations Application Note 450799 (1999)
[8] Microchip, PS501 Single Chip Field Reprogrammable Battery Manager, Datasheet DS21818C (2004)
[9] Dallas Semiconductors, Characterizing a Li+ cell for use with a fuel cell, Application Note 2412 (2003)
[10] Dallas Semiconductors, DS2781 1-Cell or 2-Cell Stand Alone Fuel Gauge IC, Datasheet (2006)
[11] V. Pop, H.J. Bergveld, P.H.L. Notten, P.P.L. Regtien, Smart and Accurate State-of-Charge Indication in Portable Applications, Power Electronics and Drive Systems, **1**, 262–267 (2005)
[12] V. Pop, H.J. Bergveld, P.H.L. Notten, P.P.L. Regtien, State-of-Charge Indication in Portable Applications, IEEE International Symposium on Industrial Electronics, **3**, 1007–1012 (2005)
[13] Maccor Inc., Help file Version 5.4 (2005)
[14] V. Srinivasan, J.W. Weidner, J. Newman, Hysteresis during Cycling of Nickel Hydroxide Active Material, J. of Electrochem. Soc., **148**, 969–980 (2001)
[15] G. Plett, Extended Kalman filtering for battery management systems of LiPB-based HEV battery packs: Part 1. Background, J. Power Sources, **134**, 252–261 (2004)
[16] G. Plett, Extended Kalman filtering for battery management systems of LiPB-based HEV battery packs: Part 2. Modeling and identification, J. Power Sources, **134**, 262–276 (2004)
[17] G. Plett, Extended Kalman filtering for battery management systems of LiPB-based HEV battery packs: Part 3. State and parameter estimation, J. Power Sources, **134**, 277–292 (2004)
[18] W. S. Kruijt, H.J. Bergveld, P.H.L. Notten, Electronic network model of a rechargeable Li-ion battery, Internal Philips Technical Note 265 (1998)
[19] N. Korver, State-of-Charge estimation in GSM applications, Report 001M05, University of Twente (2005)
[20] P.P.L. Regtien, F. van der Heijden, M.J. Korsten, W. Olthuis, Measurement Science for Engineers, Kogan Page Science Publisher, London (2004)

Chapter 4
Methods for measuring and modelling a battery's Electro-Motive Force

Electro-Motive Force (EMF) measurement and, modelling are described in this chapter and simulation results are presented. EMF measurement by means of linear interpolation and voltage-relaxation methods is presented in section 4.1. A new EMF prediction model is described in section 4.2. This model is needed for a correct interpretation of the EMF measurement results. The focus in section 4.3 is on EMF hysteresis, a phenomenon discovered during analysis of the measurement results. A new EMF model that also includes EMF temperature dependence and simulation results obtained with this model are presented in section 4.4, in which the simulation results are also compared with the measurement results. Section 4.5 presents concluding remarks.

4.1 EMF measurement

EMF is a battery's internal driving force for providing energy to a load. The battery voltage only equals the EMF when no current flows and the voltage has relaxed to its equilibrium value, *i.e.* the EMF.

Two EMF determination methods will be considered in this chapter: *linear interpolation* and *voltage relaxation* [1]–[3]. Sony's US18500G3 Li-ion batteries were used in all the experiments and simulations discussed in this chapter. At the time of testing the batteries were fairly new, having undergone 9 discharge/charge cycles. Table 4.1 presents the main characteristics of the US18500G3 Li-ion battery.

Table 4.1. US18500G3 Li-ion battery characteristics.

Battery	Characteristics
Chemical system	Lithium-ion
Cell Type	US18500G3
Cell diameter	at most 18.4 mm
Cell Length	at most 49.3 mm
Capacity (0.2 C-rate), typical	1180 mAh (3.0 V cut off)
Capacity (0.2 C-rate), minimum	1100 mAh (3.0 V cut off)
Cell weight	33 g

In the *linear interpolation* method the average battery voltage, calculated at the same SoC, is inferred from the battery voltages during two consecutive discharge and charge cycles at the same C-rate and temperature. The average of the charge and discharge voltages is taken in order to minimise the possible effects of overpotential and hysteresis in the EMF function. In the calculations discussed in this chapter the EMF was determined using the *linear interpolation* method, which comprises the following steps. First a battery is fully charged to 4.2 V at a constant 0.05 C-rate. At the end of the charge cycle the SoC level is defined to be 100%. The charging step is followed by a rest period of 24 hours, after which a discharge step is applied at a constant 0.05 C-rate until the battery voltage reaches 3 V. At the end of the discharge cycle the SoC level is defined to be 0%. The low C-rate value was chosen to minimise the effect of the overpotential. The long rest periods were chosen to ensure that a new cycle would always start in the equilibrium state. This way the effect of a not-fully-relaxed voltage is eliminated from the EMF determination. Fig. 4.1 shows the EMF curve obtained at 25°C using the linear interpolation method.

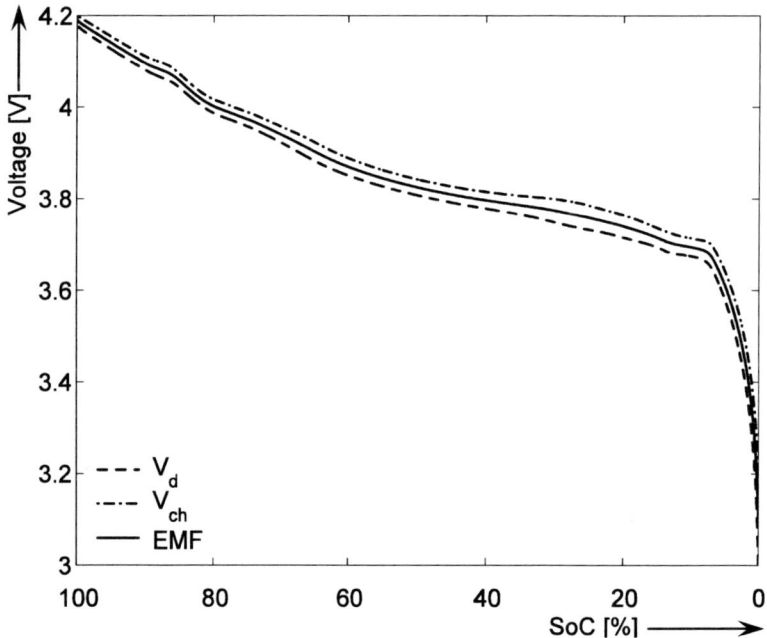

Fig. 4.1. EMF curve (EMF) obtained with the linear interpolation method at 25°C. V_{ch} represents the battery voltage measured during charging and V_d that measured during discharging. The horizontal axis shows the SoC [%] normalised to maximum capacity.

As during (dis)charging, the battery voltage was found to be (lower) higher than the EMF (see Fig. 4.1). The linear interpolation method is based on the assumption that at a given SoC the overpotential is symmetrical during discharge and charge cycles. As the overpotential is time- and SoC-dependent [1], the starting point in time and SoC of the experiment during the charge and discharge cycles should be the same to obtain a symmetrical overpotential. Fig. 4.2 shows the charge and discharge overpotentials determined as a difference between the EMF obtained by means of voltage-relaxation, as will be described below, and the charge/ discharge voltage.

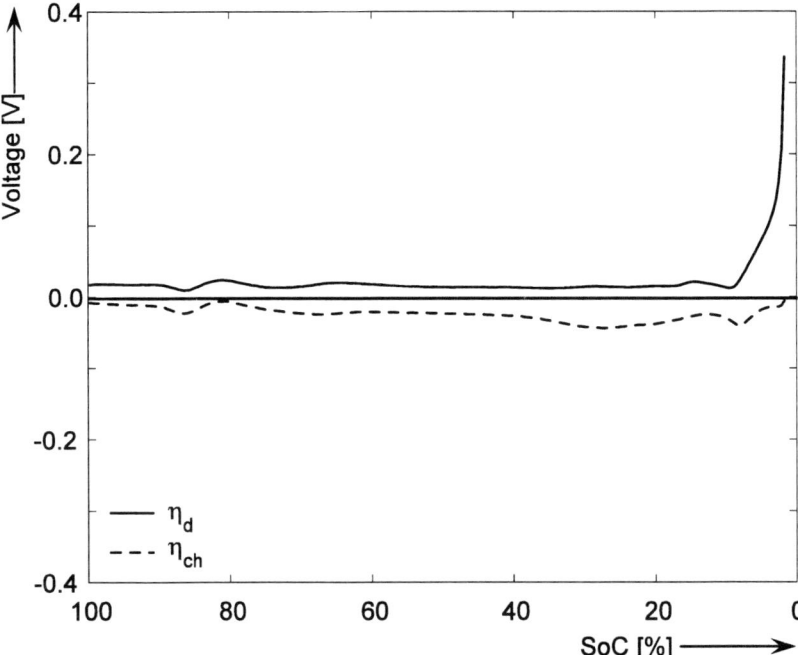

Fig. 4.2. The charge (η_{ch}) and discharge (η_d) overpotentials obtained as the difference between the EMF and charge/discharge voltage at 0.05 C-rate and 25°C, the same C-rate current and T conditions as used in the linear interpolation method. The EMF was experimentally determined by means of *voltage relaxation* [1]. The horizontal axis shows the SoC [%] normalised to maximum capacity.

It can be concluded from Fig. 4.2 that the charge and discharge overpotentials are not symmetrical. This non-symmetry is caused by a different build-up of the overpotential as a function of SoC in the charge and discharge cycles, particularly at low SoC values. More information on the overpotential will be given in chapter 5. *Linear interpolation* is therefore not a preferred method for determining the EMF.

In another known method the EMF is determined via *voltage relaxation*. The battery voltage will relax to the EMF value after current interruption. This may take a long time, especially when a battery is almost empty, at low temperatures and after a high discharge current rate. Fig. 4.3 illustrates what happens to the battery voltage after a discharge step at 0.25 C-rate current and 5°C. In order to guide the eye, the battery voltage during the relaxation period has been plotted as a function of logarithm of the relaxation time in Fig. 4.4. As can be seen in Figs. 4.3 and 4.4, the Open-Circuit Voltage (OCV) does not coincide with the battery's EMF voltage for the greater part of the relaxation process. The value of the OCV changes from 3 V after the current interruption to about 3.748 V after 600 minutes. It can be observed in Fig. 4.4 that the OCV becomes constant after about 100 minutes. The voltage after 30 minutes differs by approximately 15 mV from the voltage after 600 minutes. This means that when in this example it is assumed that the equilibrium state is reached after 30 minutes, the inaccuracy in SoC indication will be about 4.2%; see Fig. 4.3.

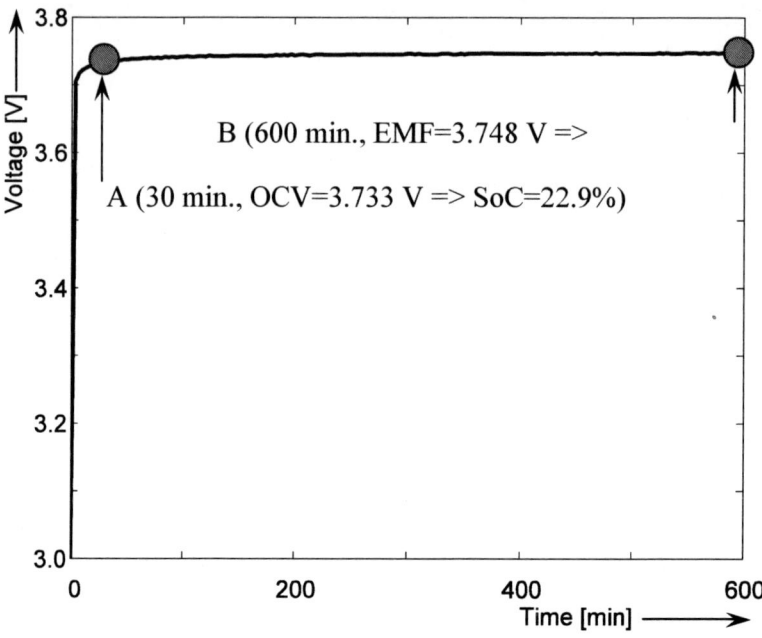

Fig. 4.3. Voltage-relaxation after a discharge current step of 0.25 C-rate at 0% SoC and 5°C. The horizontal axis shows the time in [min.].

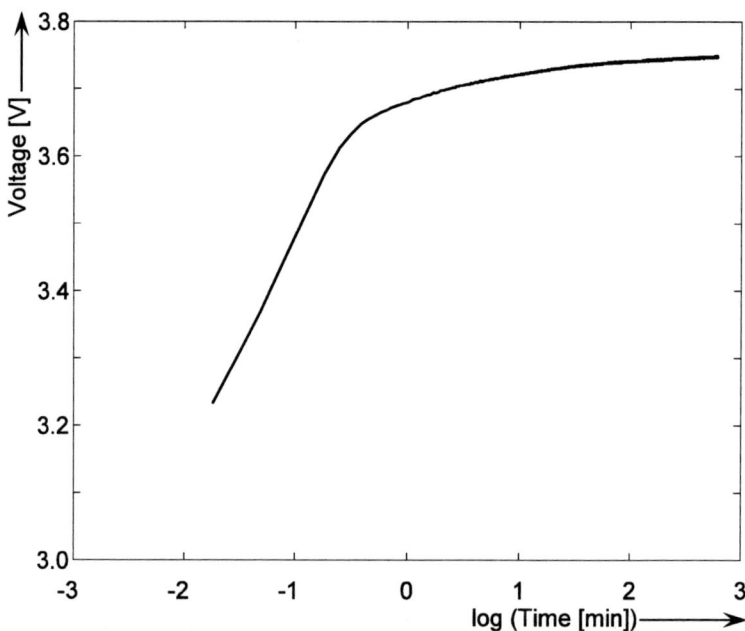

Fig. 4.4. Voltage-relaxation after a discharge current step of 0.25 C-rate at 0% SoC and 5°C. The horizontal axis shows the logarithm of the time $log_{10}(Time[min.])$.

In the experiment discussed in this chapter the voltage-relaxation method was applied as follows. First, the battery was charged at a 0.1 C-rate in 25 steps. For each step in the experiment a limit of 4.2 V in voltage and 50 mAh in capacity was assumed. Four "Deep-Charge" steps at 0.05 C-rate were then applied. At the end of these steps the SoC level was defined to be 100%. Each charge step was followed by a rest period. The rest period was chosen as a function of SoC, i.e. at low SoCs long relaxation times of 24 hours were used and at higher SoCs shorter relaxation times of 8 hours. After charging, 25 discharge steps of 50 mAh each were applied at a 0.1 C-rate and a voltage limit of 3 V. Four "Deep-Discharge" steps at 0.05 C-rate were then applied. At the end of these steps the SoC level was defined to be 0%. Each discharge step was followed by a rest period chosen in the same way as in the case of the charge steps. The low C-rate was chosen to obtain an equilibrium voltage faster. The experiment was repeated at different C-rates and temperatures, each time yielding 29 points of the EMF-SoC curve.

The 29 measured EMF points were fitted using a newly developed method in which the shape of the curve is also taken into consideration. In this way, 20000 fitted points were obtained, which yielded measured EMF values for each 0.005% increment in SoC value. Fig. 4.5 illustrates the EMF curve obtained with the voltage-relaxation method during the discharge cycles.

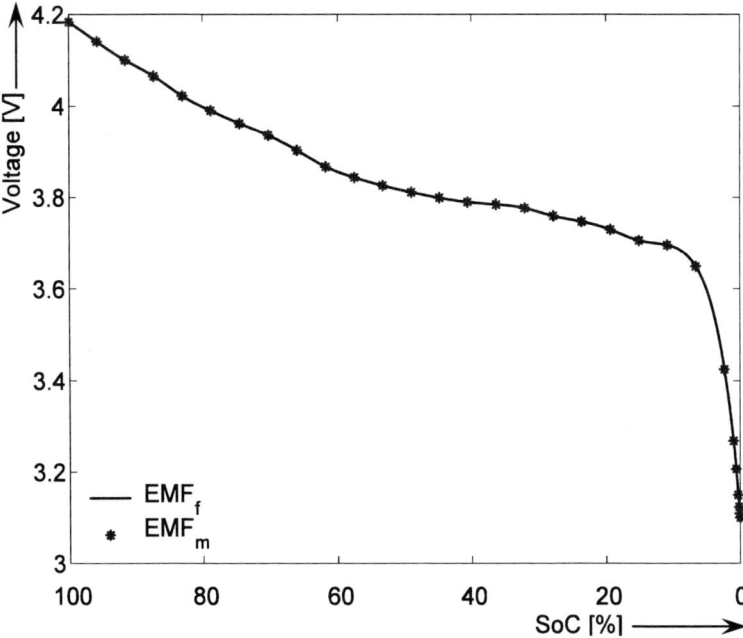

Fig. 4.5. Measured EMF data points (EMF_m) and the fitted EMF curve (EMF_f) obtained with the voltage-relaxation method during the discharge cycles. The horizontal axis shows the SoC [%] normalised to maximum capacity.

Fig. 4.6 presents the EMF obtained with the voltage-relaxation method as a function of SoC at three temperatures, i.e. 0, 25 and 45°C, during the discharge cycles. In order to guide the eye, the differences between the EMF values obtained at different temperatures have been indicated in Fig. 4.7.

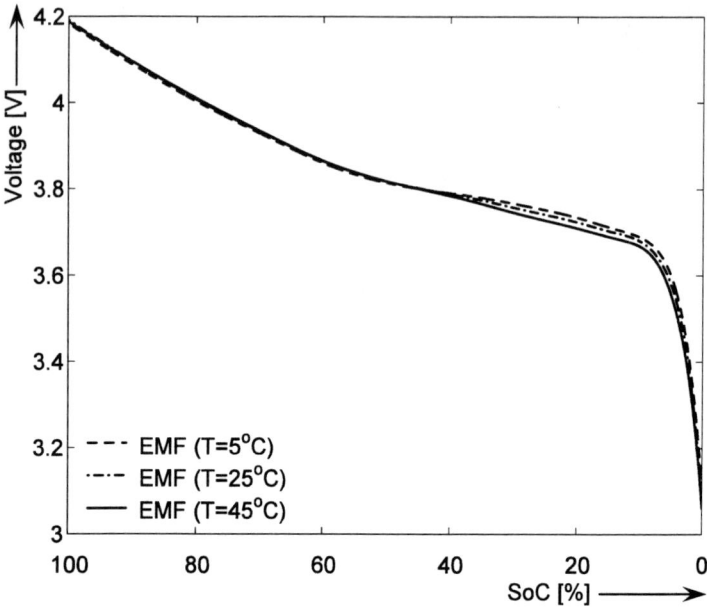

Fig. 4.6. EMF measured by means of voltage-relaxation during the discharge cycles as a function of SoC at three temperatures. The horizontal axis shows the SoC [%] normalised to maximum capacity.

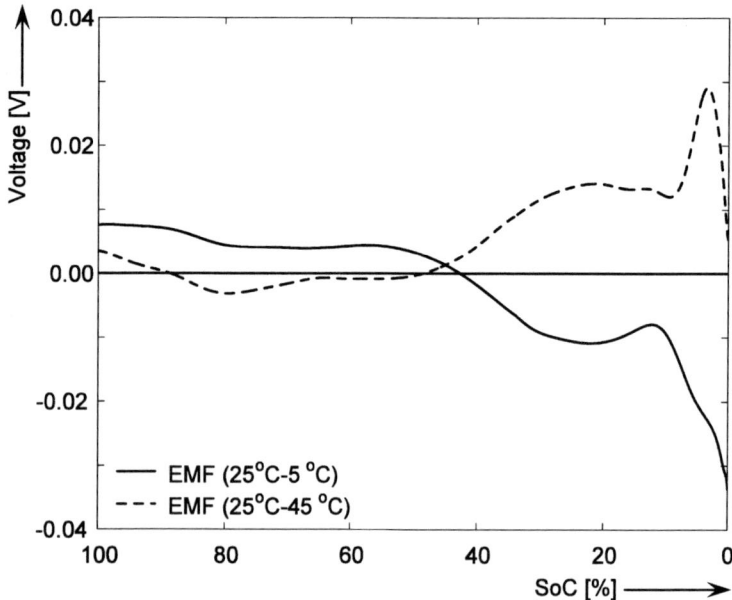

Fig. 4.7. EMF difference as a function of SoC at three temperatures. The horizontal axis shows the SoC [%] normalised to maximum capacity.

It can be concluded that the maximum difference between the EMF measured at 25°C and that measured at 45°C is about 29 mV at around 4% SoC. This means that when the temperature effect is not taken into consideration and the EMF curve is modelled only at 25°C, the SoC indication system based on the EMF will yield an SoC value of 3.6% at 45°C, whereas the actual SoC value will be 4.1%. The inaccuracy, calculated as the difference between the actual SoC value measured at 25°C and the SoC value measured at 45°C, will be 0.5% [4]. This effect will be more pronounced at low temperatures and in the flat region of the EMF-SoC curve, where even minor differences in the EMF will cause substantial errors in SoC.

In summary, the voltage-relaxation method is recommended for EMF determination. For accurate SoC determination, the EMF dependence on the temperature effect should moreover be taken into consideration.

4.2 Voltage prediction

This section presents a new mathematical model for the Li-ion battery voltage-relaxation process. Using this model the equilibrium voltage, *i.e.* EMF, can be predicted accurately (*i.e.* with less than 1% SoC error) and within a short time (*i.e.* less than 6 minutes). It should be noted that the model parameters will be adapted on-line on the basis of battery voltage measurements, so the overall concept can be used for any other battery chemistry. An important advantage of the voltage-relaxation model is that it can speed up SoC indication on the basis of the EMF. This is necessary to calibrate for drifts resulting from the Coulomb counting and to improve the maximum capacity adaptation capabilities [1].

4.2.1 Equilibrium detection

It is important that the system is able to reach the equilibrium state, because in this state less calculation effort is required, which will mean that the SoC indication system will consume less power. If the system is in equilibrium the SoC value calculated on the basis of the EMF-SoC relationship will moreover be more accurate and the maximum capacity adaptation capability can be improved [1]. If the system reaches the equilibrium stage before the battery voltage has stabilised the voltage will be higher (when current has been interrupted after a charge step) or lower (when current has been interrupted after a discharge step) than the EMF, leading to a too high (return from charge step) or too low (return from discharge step) predicted SoC value. When such an incorrect SoC value is used for calibration, the system's accuracy will be compromised. The same holds for the updating of the maximum capacity. The latter problem is discussed in [1], [3], in relation to an experiment in which the maximum capacity was updated after applying a discharge step. Since the battery voltage had not fully relaxed by the time that the algorithm returned to the equilibrium state, the SoC predicted on return to the equilibrium state after the application of the discharge step was too low, and the resulting calculated maximum capacity was also too low. There are several methods for detecting the equilibrium state. They will be briefly introduced in this section.

Fixed time to wait

A simple method is to wait for a fixed length of time after current interruption and to assume that the battery voltage is stable after this time. In this situation the longest possible, *i.e.* worst-case, relaxation time must be used to be sure that the battery is indeed in equilibrium. In the case of low SoC and temperature values this may take a long time; see *e.g.* Figure 4.3. Such long rest periods are very rarely used in portable devices. The waiting time can be also chosen as a function of *e.g.* SoC and temperature but even then false entries into the equilibrium state are likely to occur due to *e.g.* spread between batteries. The system moreover has to wait until the battery has relaxed for the EMF value to become available and calibration or updating of maximum capacity becomes possible.

Threshold for the change in battery voltage with time (dV/dt)

In order to allow the algorithm to change to the equilibrium state, the condition of a stable voltage has to be met. The voltage change in time, *i.e.* the derivative dV/dt, can be used to determine this: when dV/dt is below a certain threshold value the voltage may be assumed to be stable and the battery to be in equilibrium. Under normal conditions a battery will however never reach a fully relaxed state because a certain small current will always be present (*e.g.* in a mobile phone application the standby current). As already mentioned in chapter 3, due to the EMF and overpotential dependence on the SoC, it is difficult to distinguish between a relaxed and a non-relaxed battery voltage by using only dV/dt measurements. If the same threshold value is used for all SoC values to detect equilibrium (dV/dt < threshold) the risk of false detections seems to be quite high. Another disadvantage of this method is that the system has to wait until the battery has fully relaxed. This solution was proposed in [1]. In [3] the claim was that the battery is checked on having a stable voltage, without specifically mentioning the method.

Voltage prediction

When a model of the voltage-relaxation process is available, the relaxation end value, *i.e.* the EMF, can be calculated on the basis of the relaxation conditions. The main advantage over using a fixed waiting time or dV/dt threshold is that the EMF value becomes available before a battery has fully relaxed. This allows the system to calibrate the SoC obtained in prolonged Coulomb counting even when a user does not leave enough time for full relaxation between successive charges or discharges. This of course imposes strong demands on the accuracy of the predicted relaxation end value.

Of the three described methods for determining whether a battery is in equilibrium the third is the most advantageous as it offers more calibration and update opportunities for the maximum battery capacity. Another two patented methods are to be found in the literature [5], [6].

4.2.2 Existing voltage-relaxation models used for voltage prediction

Two modelling methods known in the state-of-the-art for predicting voltage relaxation will be presented in this section.

Asymptotes method

The voltage-prediction method presented by Aylor *et al.* focuses on lead-acid batteries [5]. The principle of this method is that a battery's OCV recovery curve (solid line) can be approximated by two asymptotes if plotted on a semi-log scale, as shown in Fig. 4.8 (the dashed line is the first asymptote and the solid line is the second asymptote). A new variable X is introduced instead of t (time) in order to obtain a linear equation, with X being the logarithm of time.

$$t\ [min.] = 10^X \qquad (4.1)$$

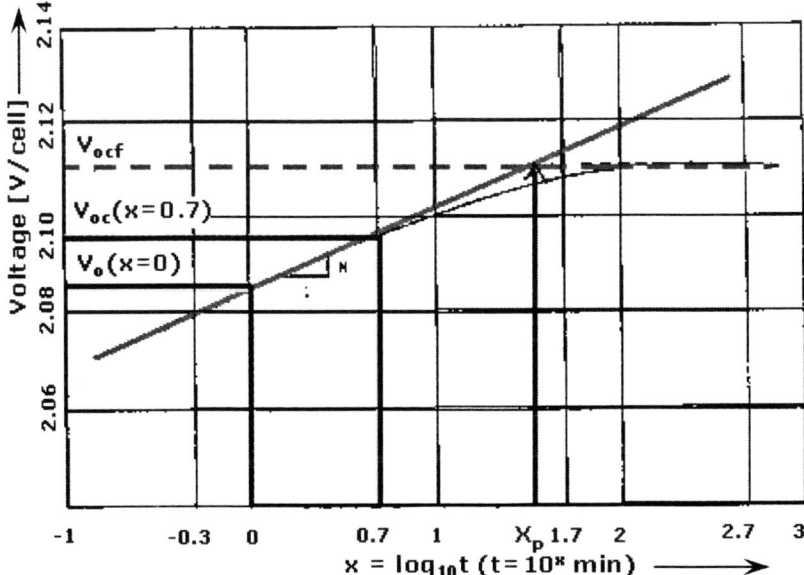

Fig. 4.8. Open Circuit Voltage as a function of X (the logarithm of time).

From Fig. 4.8 it can be inferred that

$$M = \frac{V_{ocf} - V_o}{X_p - 0} \Rightarrow V_{ocf} = MX_p + V_o \qquad (4.2)$$

where M represents the slope of the second asymptote (the solid line in Fig. 4.8), V_{ocf} denotes the fully stabilised Open-Circuit Voltage, X_p denotes the value of X at the asymptotes' intersection and V_o equals the battery voltage at $t=1$min. ($X = 0$).

After performing calculations for different types of lead-acid cells [5] concludes that an average value of 1.64 for X_p yields the greatest accuracy in voltage prediction. Therefore, according to Eq. (4.2)

$$V_{ocf} = M\ 1.64 + V_0 \qquad (4.3)$$

For M inferred from $V_{oc} = V\ (t = 6.6$ min.$)$ and $V_0 = V\ (t =1$ min.$)$ Eq. (4.3) becomes:

$$V_{ocf} = 2(V_{oc} - V_0) + V_0 = 2V_{oc} - V_0 \qquad (4.4)$$

Note that this method makes use of a fixed parameter obtained in laboratory experiments, *i.e.* $X_p=1.64$, to predict the relaxation end voltage. If this parameter X_p is not updated, the system will become progressively less accurate in predicting the voltage-relaxation end value, *i.e.* the EMF, as a battery ages and its voltage-relaxation behaviour changes. Results obtained with this voltage-relaxation method applied to a Li-ion battery will be presented in section 4.2.5.

dV/dt and temperature combined method

In [6] Hoenig *et al.* present a voltage-prediction model based on the measured battery OCV, the change in measured battery OCV over time (dOCV/dt), and the temperature measured at any time during the voltage-relaxation process. The invention is described in relation to lead-acid storage batteries having sufficient cells to produce a rated voltage of 24 V. Applicability to other battery chemistries is however also claimed.

In order to find the correct parameters for the system, a battery is charged and discharged in fixed steps, starting with an empty battery. Each time the charge or discharge current is interrupted three parameters are measured and recorded as data points until the battery voltage/OCV has stabilised. These parameters are the battery voltage (OCV), the rate of change of the battery OCV and the temperature. The battery OCV and temperature are measured instantaneously, while the rate of change of OCV is measured over a predetermined period, *i.e.* 30 seconds. These measurements performed in a laboratory using the selected lead-acid battery led to two equations for the predicted relaxation end voltage, depending on the battery voltage. The voltages were translated into single-cell voltages. Since the average single-cell voltage of lead-acid batteries is 2 V, 12 cells were used in the 24 V battery.

The aforementioned two equations are:

for battery Open-Circuit Voltages below (2.08 V) (number of cells)

$$V_p = 1.165 OCV + 6.95\, dOCV/dt - 0.167\, T - 0.95 \qquad (4.5)$$

and for battery Open-Circuit Voltages above (2.08 V) (number of cells)

$$V_p = 1.027\, OCV + 9.288\, dOCV/dt - 0.0197\, T - 0.56 \qquad (4.6)$$

where V_p denotes the predicted relaxation end voltage, *OCV* denotes the measured battery Open-Circuit Voltage and *dOCV/dt* represents the change in battery *OCV* over time. *T* denotes the battery temperature in °C.

The transfer OCV between Eqs. (4.5) and (4.6) can be determined via a rough OCV prediction or measurement. According to the patent, the transfer will occur around a battery SoC of 75%. It is claimed that the exact definition of this transfer point is not important, since the inaccuracy introduced by use of a wrong equation would be minimal in the region around 75% SOC. In the case described in the patent for the lead-acid battery rated at 24 V the value of 25 V is used as the decision point. This value can be translated to 2.08 V multiplied by the number of cells, *i.e.* 12. Note again that Eqs. (4.5) and (4.6) make use of fixed parameters obtained in a laboratory experiment. When these parameters are stored in the system and the battery characteristics change, which will always happen due to ageing, the accuracy of the predicted voltage-relaxation end values will decrease over time. Results obtained with this voltage-relaxation method using a Li-ion battery will be presented in the next sections.

In summary, the main problem with the SoC indication method patented in [3] is accurate determination of a battery's equilibrium. Of the available methods for determining a battery's equilibrium voltage after relaxation that based on a voltage-prediction model seems the most attractive. The methods currently known in the literature however make use of fixed parameters in the prediction of the voltage-relaxation end value, which is a disadvantage, because changing battery characteristics due to aging will lead to decreasing prediction accuracy over time [5], [6].

4.2.3 A new voltage-relaxation model

The method described in this section is a voltage-prediction method without the need to store parameters beforehand. Instead, the voltage-relaxation end value is determined on the basis of the measured first part of a voltage-relaxation curve and mathematical optimisation/fitting of a function to this measured part of the relaxation curve. In addition to the unknown voltage-relaxation end value, the function contains three more parameters that are also found by fitting. This means that these parameters are updated for each individual situation, without the need to store values beforehand. The advantages are *(i)* that the EMF can be predicted with enough accuracy in the first few minutes after current interruption and *(ii)* that no previously stored parameters are used as in the methods described in [5] and [6]. The first advantage improves the SoC determination by offering more calibration opportunities and solving the problem of the inability to determine a battery's exact equilibrium. It will also improve any SoC indication system based on the EMF method. The second advantage makes the method more suitable for dealing with battery aging than the voltage-prediction methods described in [5] and [6].

The model

The method can be described as follows. At the beginning of the relaxation process a battery's actual open-circuit voltage does not coincide with the EMF. The cause of this difference is the overpotential built up during the preceding (dis)charge period. This overpotential makes the battery voltage during the (dis)charge process deviate from the EMF. The overpotential build-up is caused by various electrochemical processes that occur in the battery, such as Li^+-ion diffusion in both electrodes, diffusion and migration of Li^+ and other ions in the electrolyte, Butler-Volmer kinetic limitations at the surface of the electrodes, *etc.* [1]. The relaxation process is consequently generally speaking a complex function involving several factors, such as the SoC, amount of charge added or removed from the battery, temperature and aging. This large number of dependencies makes it difficult to predict the EMF at the very beginning of the relaxation period. For this reason, the relaxation process should be observed for a certain length of time to arrive at an accurate prediction of the final EMF value.

Denote $t_1, \ldots, t_N$ [s] as the moments in time at which the battery voltages $V_{t_1}, \ldots, V_{t_N}$ [V] are observed during the relaxation process. This implies that the first part of the relaxation curve is sampled at N sample points (t_i, V_{t_i}). The general model for the voltage-relaxation process proposed in this section is given by the following equation [7]:

$$V_t = V_\infty - \frac{\Gamma \gamma}{t^\alpha \log^\delta(t)} e^{\varepsilon_t/2} \qquad (4.7)$$

where parameters $\gamma > 0, \alpha > 0, \delta > 0$ are the rate-determining constants, Γ is equal to $+1$ if the voltage is increasing (relaxation after a discharge step) and equal to -1 if the voltage is decreasing (relaxation after a charge step), V_∞ is the final relaxation voltage (an asymptotical value, i.e. EMF), V_t is the relaxation voltage at time t, $\log^\delta(t)$ is the natural logarithm (with base e) of time [s] to the power of parameter δ and ε_t is a random error term. Note that the model presented in Eq. (4.7) assumes an exponential multiplicative error structure[1].

The parameters V_∞, α, δ, and γ can be estimated by applying the concentrated Ordinary Least Squares (OLS) scheme. Therefore, Eq. 4.7 can be re-written by taking squared values:

$$(\Gamma \frac{\gamma}{t^\alpha \log^\delta(t)})^2 = (V_\infty - V_t)^2 e^{-\varepsilon_t} \qquad (4.8)$$

Taking logarithm reduces Eq. (4.8) to a usual linear regression model:

$$\log(V_\infty - V_t)^2 = C + A\log(t) + D\log(\log(t)) + \varepsilon_t \qquad (4.9)$$

where $C = 2\log(\gamma)$, $A = -2\alpha$ and $D = -2\delta$. For each fixed value of V_∞ Eq. (4.9) can be written as the usual regression model

$$y_t = x_t' \beta + \varepsilon_t, \quad t = t_1, \ldots, t_N \qquad (4.10)$$

where $y_{t_i} = \log(V_\infty - V_{t_i})^2$, $x_{t_i} = (1, \log(t_i), \log(\log(t_i)))'$, $\beta = (C, A, D)'$, $i=1,\ldots,N$. In matrix notations it is written as

---
[1] Note that the model described by Eq. (4.7) can be formally written in an additive form, i.e. $V_t = V_\infty - \frac{\Gamma \gamma}{t^\alpha \log^\delta(t)} + \varsigma_t$ while the error term ς_t is dependent on V_t.

Methods for measuring and modelling a battery's Electro-Motive Force

$$y = X\beta + \varepsilon \qquad (4.11)$$

where $y = (y_1, \ldots, y_N)'$, $\varepsilon = (\varepsilon_1, \ldots, \varepsilon_N)'$ and $X = (x_1, \ldots, x_N)'$ is a matrix of regressors based on N available time instants t_i, i.e. the columns in this matrix are formed by ones, $\log(t_i)$-values and $\log(\log(t_i))$-values. After the initial OLS estimator has been obtained for β, i.e. $\hat{\beta} = (\hat{C}, \hat{A}, \hat{D})'$, where $\hat{\beta} = (X'X)^{-1}X'y$ based on an initial guess for V_∞ used to calculate y, $\|y - X\hat{\beta}\|^2$ is calculated and minimised with respect to V_∞ (where the OLS estimator $\hat{\beta}$ is re-calculated each time V_∞ is changed). This leads to an implementation that is simpler than minimising $\|y - X\hat{\beta}\|^2$ with respect to V_∞ and β jointly. The initial guess of V_∞ can be obtained for example by adding 100 mV to the first voltage sample in the case of relaxation after a discharge step, or by subtracting 100 mV in the case of relaxation after a charge step. Finally, the parameters in the original model of Eq. (4.7) will be recovered as $\hat{\gamma} = \exp(\hat{C}/2)$, $\hat{\alpha} = -\hat{A}/2$, $\hat{\delta} = -\hat{D}/2$. In summary, the voltage-prediction model of Eq. (4.7) is fitted to the measured relaxation curve described by sample points (t_i, V_{t_i}), $i=1,\ldots,N$. This yields a predicted relaxation voltage curve and a predicted relaxation end voltage V_∞. Since there are 4 unknown parameters in the model of Eq. (4.7) (V_∞, α, δ, γ) at least 4 sample points are needed to solve the set of equations, i.e. $N \geq 4$.

The final objective of the voltage-relaxation model is to offer SoC indication accuracy that is better than that achieved on the basis of battery OCV or by means of Coulomb counting. In this book an SoC indication accuracy calculated on the basis of the predicted EMF voltage of within 1% SoC is considered acceptable.

4.2.4 Implementation aspects of the voltage-relaxation model

To test the accuracy of the voltage-prediction model, a Li-ion US18500G3 battery (rated capacity 1100 mAh) was charged and discharged using a Maccor battery tester in steps of 50 mAh at different C-rates (0.05, 0.1, 0.25 and 0.5 C-rate) and different temperatures (0, 5, 25 and 45°C). The voltage-relaxation measurements showed that the vertical part of the relaxation curve in the first moments of relaxation may yield substantial inaccuracies in the predicted end-voltage value. To obtain an optimum value for the first sample time t_1 that would minimise the error in the predicted voltage, roughly 500 relaxation curves obtained with the Maccor battery tester as described above were simulated with the aid of the model of Eq. (4.7) using MATLAB. It was found that at least the first 0.5 minute after current interruption must be ignored. Voltage samples intended for fitting the model of Eq. (4.7) should be taken after this period of time.

Fig. 4.9 illustrates the voltage relaxation process after application of a charge step. The figure shows the battery voltage after a charge step at a 0.5 C-rate and 25°C.

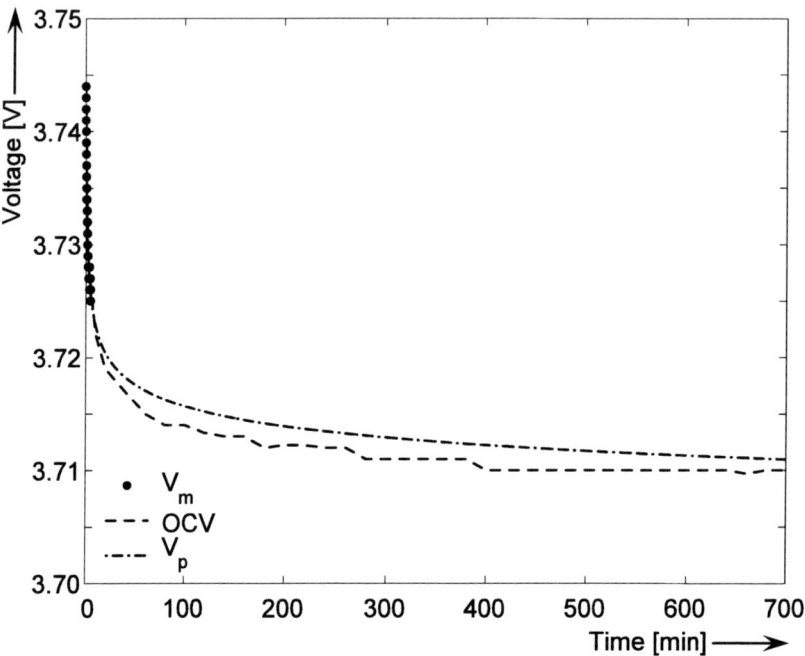

Fig. 4.9. Battery OCV relaxation after a charge step at 0.5 C-rate and 25°C. The data points measured in the first 5 minutes of the relaxation process (V_m) are used for the voltage prediction (V_p). The horizontal axis shows the time in [min.].

The voltage eventually relaxes to a stable value of 3.710 V, *i.e.* the EMF corresponding to this SoC value. After 5 minutes of relaxation, the OCV (V_m) still differ 15 mV from the end value. This means that if the algorithm were to return to the equilibrium state after these 5 minutes, the battery voltage would still differs 15 mV from the actual EMF, which would in this case correspond to a 2.5% SoC error. Returning to the equilibrium state after 60 minutes would in this case lead to an acceptable SoC error of less than 1%. This also means that in cases like the one illustrated in Fig. 4.9, a system cannot be calibrated until it has returned to the equilibrium state after 60 minutes. If a user starts recharging a battery before these 60 minutes have elapsed, the calibration opportunity will be lost. In summary, it is crucial to accurately determine when the battery voltage has stabilised after application of a charge/discharge step, and this may take a long time. It follows from Fig. 4.9 that the SoC calculated via voltage prediction after five minutes of rest is very close to the final SoC value calculated on the basis of the EMF voltage. In this example the calculated SoC inaccuracy is –0.2% SoC.

In order to further investigate the accuracy of the voltage-prediction model several measurements were performed in a laboratory set-up. In these measurements a battery was charged and discharged and the end voltage after current interruption was estimated in real-time using the voltage model of Eq. (4.7). The model's accuracy was again determined by comparing the SoC calculated with the aid of the EMF curve on the basis of the predicted EMF voltage using the model of Eq. (4.7) with an SoC calculated on the basis of a final EMF value obtained in the laboratory measurements, *i.e.* the final stabilised voltage after a long

Methods for measuring and modelling a battery's Electro-Motive Force 77

period of relaxation. The first 0.5 minute of relaxation were ignored in formulating the model of Eq. (4.7). The total relaxation period was chosen so that the battery would be able to relax completely under all conditions.

In addition, the SoC was also determined using the EMF curve based on the instantaneous OCV values of the battery voltage during relaxation. As before, the error in this calculated SoC can be calculated by comparing it with the SoC based on the final EMF value measured at the battery terminals after a long relaxation time. The latter error gives an indication of the magnitude of the error that would occur when assuming the battery is in equilibrium after a fixed relaxation time. The SoC errors obtained when using the voltage-relaxation model of Eq. (4.7) or the instantaneous OCV value obtained for a discharge at 0.25 C-rate and 5°C are shown in Fig. 4.10.

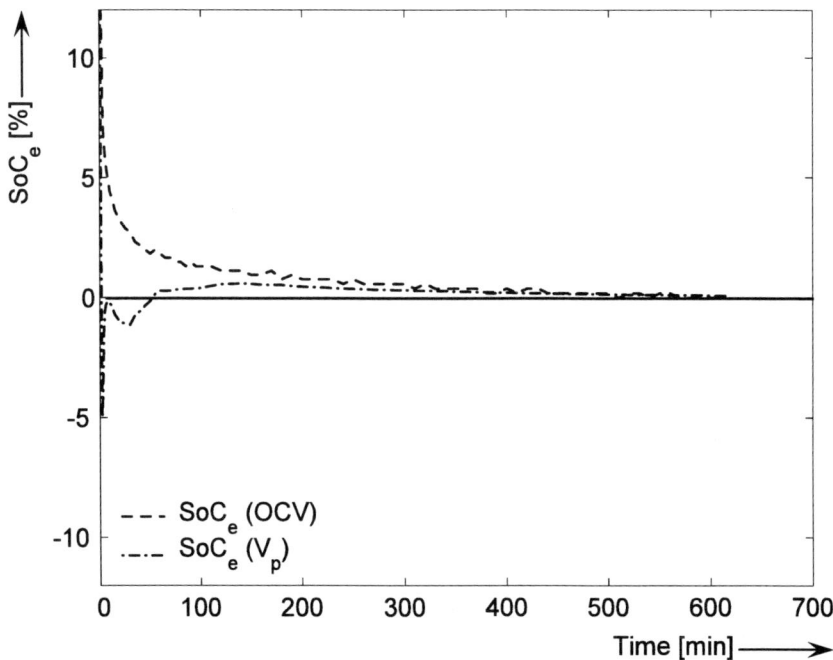

Fig. 4.10. Error in SoC calculated on the basis of a predicted voltage ($SoC_e(V_p)$) obtained using the model of Eq. (4.7) compared with an error in SoC calculated on the basis of the instantaneous OCV ($SoC_e(OCV)$) considered after a discharge step at 0.25 C-rate and 5°C. The horizontal axis shows the time in [min.].

Fig. 4.10 shows that the error in the SoC based on the voltage prediction ($SoC_e(V_p)$) is about 0.62% after five minutes of relaxation, whereas the SoC error obtained when using the instantaneous OCV value $SoC_e(OCV)$ is about 6.16% at that time. An SoC error $SoC_e(OCV)$ of 0.6% is obtained only after a relaxation period of 260 minutes. From this it can be concluded that voltage prediction results in better accuracy after five minutes than the battery's OCV considered after five minutes, and it yields the same accuracy as obtained when a battery's OCV is considered after 250 minutes of relaxation. The "speed" of the system based on EMF prediction and the voltage-relaxation model of Eq. (4.7) is improved 50 times in this situation (i.e. 250/5 = 50). This means that, by considering only the first five minutes of relaxation and ignoring the first half minute, the SoC can be predicted on the basis of the EMF curve with the same accuracy as obtained when using a fixed relaxation time of 250 minutes. Moreover, it can be concluded from Fig. 4.10 that for the first 400 minutes the SoC values obtained on the basis of the predicted voltage are more accurate than the SoC values obtained on the basis of the battery's OCV. After this point the two SoC values are more or less the same. Other measurement results obtained after interrupting a discharge at 0.25 C-rate and 25 and 45°C are presented in Figs. 4.11 and 4.12.

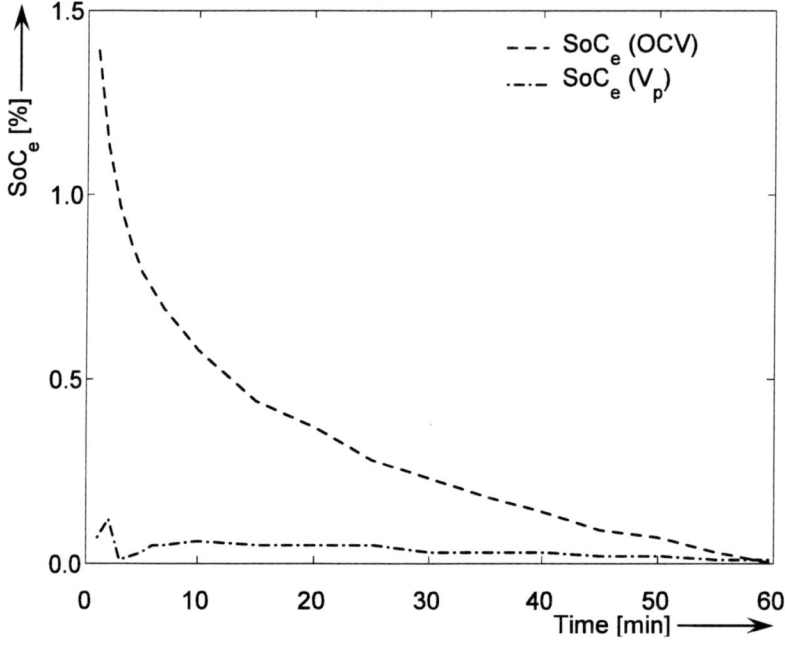

Fig. 4.11. Same as Fig. 4.10, only now at 25°C.

In both Fig. 4.11 and Fig. 4.12 the error $SoC_e(V_p)$ obtained when using voltage prediction by fitting the model of Eq. (4.7) to the first part of the measured relaxation curve and using the predicted voltage in the EMF curve is smaller than the error $SoC_e(OCV)$ obtained by filling in the instantaneous OCV value in the EMF curve. Fig. 4.12 for example shows that voltage prediction leads to an error $SoC_e(V_p)$ of 0.3% when the first two minutes of the relaxation curve are considered and the first 0.5 minute is ignored in the model fitting process, whereas $SoC_e(OCV)$ is 0.83% at that time. In this example the SoC obtained via voltage prediction using Eq. (4.7) is more accurate throughout the entire relaxation time. So from the measurements presented so far it can be concluded that the voltage-prediction model yields results that are better than those obtained by just considering an OCV value, and that an SoC error of less than 1% will usually be obtained after five minutes of relaxation.

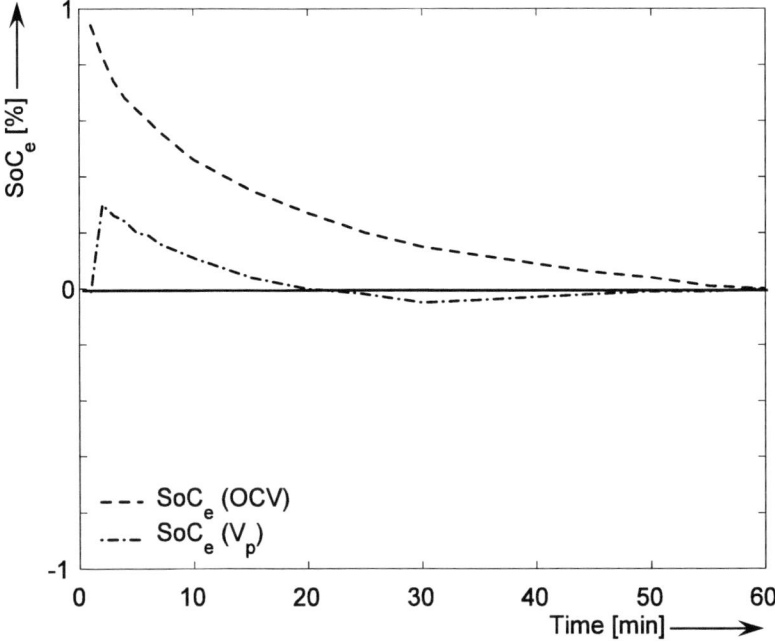

Fig. 4.12. Same as Fig. 4.10, only now at 45°C.

The same conclusions can be drawn from Figs. 4.13–4.15, in which the same curves are compared after the application of a charge step at 0.5 C-rate at various temperatures. In this case, $SoC_e(V_p)$ was obtained by fitting the model of Eq. (4.7) to the relaxation curve and ignoring the first 0.5 minute.

See for example Fig. 4.13 at a time of around four minutes. At this time, voltage prediction yields an $SoC_e(V_p)$ error of 1.3%, whereas the $SoC_e(OCV)$ error is 7.3%. An $SoC_e(OCV)$ error of around 1.3% is obtained after a rest period of 70 minutes. So voltage prediction yields greater accuracy than the battery's OCV considered after four minutes and the same accuracy as the OCV considered after 70 minutes of relaxation. The "speed" of the system based on voltage-relaxation model is improved 17 times in this situation. Fig. 4.13 also shows that for the first 110 minutes the SoC values obtained on the basis of the predicted voltage are more accurate than the SoC values obtained on the basis of the battery's OCV. After this point the two SoC values are more or less the same.

It can be concluded from Figs. 4.9–4.15 that the SoC calculated on the basis of the predicted voltage after five minutes of relaxation at different SoC values, charge/discharge rates and temperatures yields an SoC prediction error of less than 1%. A considerable gain in speed is achieved in comparison with using a fixed relaxation time, which means that after five minutes of relaxation the SoC of the battery can already be predicted with an error of less than 1% on the basis of the EMF curve, while the battery voltage has not yet relaxed to the EMF value.

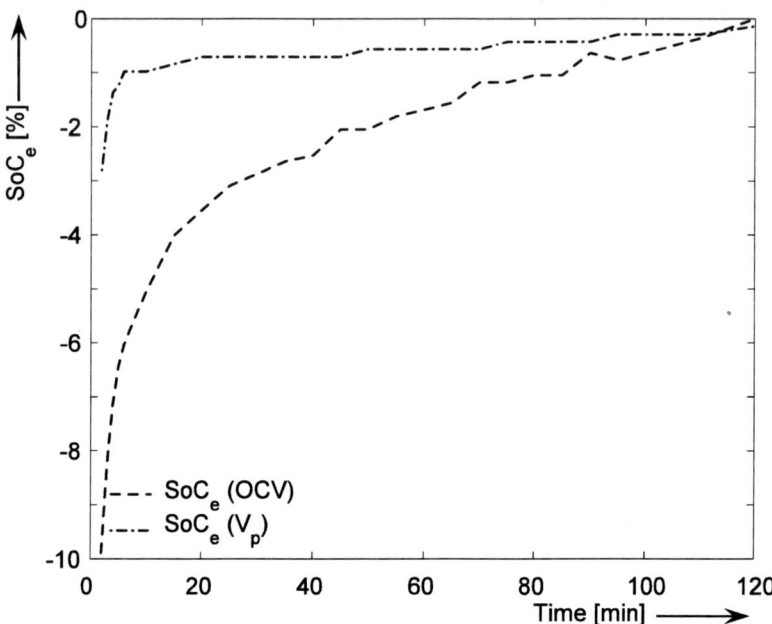

Fig. 4.13. Same as Fig. 4.10, only now after interruption of a charge step at 0.5 C-rate and 5°C.

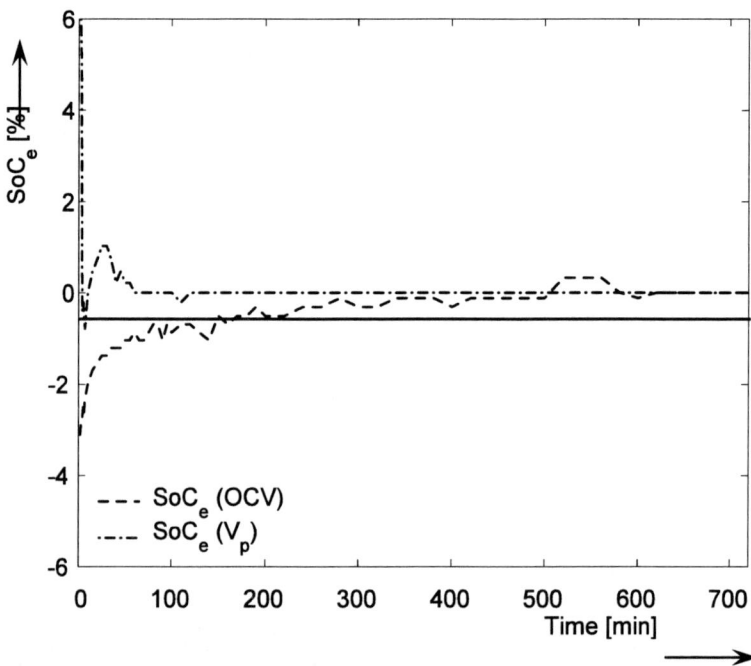

Fig. 4.14. Same as Fig. 4.10, only now after interruption of a charge step at 0.5 C-rate and 25°C.

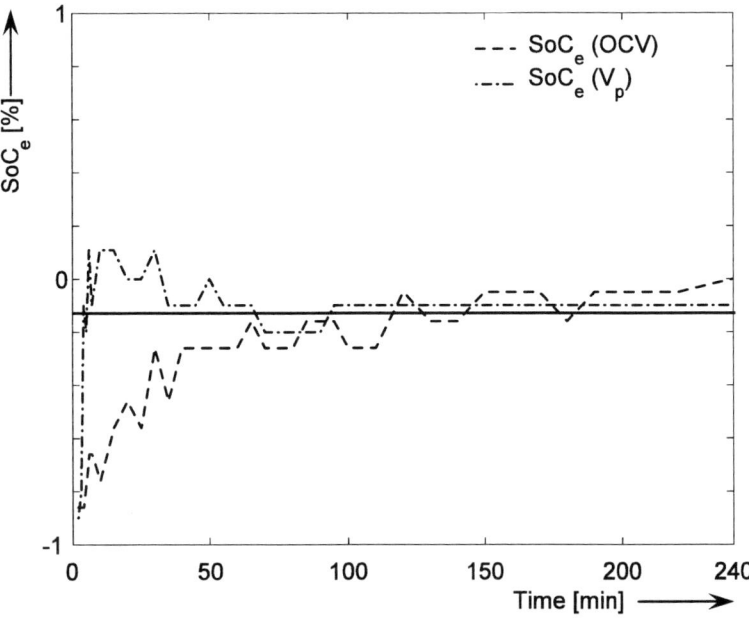

Fig. 4.15. Same as Fig. 4.10, only now after interruption of a charge step at 0.5 C-rate and 45°C.

4.2.5 Comparison of results obtained with the different voltage-relaxation models

The method of fitting the voltage-prediction model of Eq. (4.7) on-line to the first part of the relaxation curve introduced in this chapter was compared with the prior-art voltage-prediction methods presented in [5] and [6]. For the asymptotes system described in [5] the first voltage sample was taken at 1 minute (X=0) and the second at 6.6 minutes (X=0.82). The same parameters as proposed in [6] were used for the system based on OCV, $dOCV/dt$ and temperature, as well as OCV and $dOCV/dt$ values at 6.6 minutes. The results were also compared with the results of the relaxation experiments used to draw Figs. 4.9 and 4.10. The results of the comparison are summarised in Table 4.2.

Table 4.2. Comparison of results obtained with different voltage-prediction models

Model	Step	EMF [V]	V_p [V]	SoC_t [%]	SoC_p [%]	SoC_e [%]
Asymtotes [5]	d	3.748	3.753	20.19	21.11	−0.92
	ch	3.710	3.716	13.7	14.7	−1
Combined model [6]	d	3.748	2.56	20.19	0	20.19
	ch	3.710	0.7	13.7	0	13.7
New model (Eq. (4.7))	d	3.748	3.749	20.19	20.37	−0.19
	ch	3.710	3.711	13.7	13.9	−0.2

Columns one and two of Table 4.2 give the model's name and the type of previous step, *i.e.* discharge (*d*) or charge (*ch*). The equilibrium and the predicted voltage (V_p) values in [V] of each of the three models are given in columns three and four respectively. Columns five and six denote the SoC indication calculated on the basis of EMF, (SoC_t), and V_p, (SoC_p), respectively. Column seven denotes the error in SoC calculated as the difference between the actual and the predicted SoC value.

It can be concluded from table 4.2 that the calculated SoC error SoC_e obtained using the newly developed relaxation model after a discharge step is −0.19% SoC. An SoC error of 0.92% is obtained when using the asymptotes model described by Eq. (4.4), whereas the combined model described by Eq. (4.5) yields an error of 20.19% SoC.

Table 4.2 clearly shows that the SoC error obtained with the new method proposed in this chapter is smaller than that obtained with the two prior-art systems. The asymptotes system of reference [5] worked remarkably well in this Li-ion battery experiment. However, it is based on a fixed parameter X_p, which will be different in the case of other batteries of the same type and older batteries. In fact, the new model also uses parameters t_1 and N that describe which part of the relaxation curve is used for fitting Eq. (4.7). These parameters do not describe the actual relaxation curve but do influence the prediction accuracy. An advantage of the new model over the asymptotes method is that in addition to the relaxation end voltage, the time it takes to reach this voltage is also predicted. This time can be used for tuning parameters t_1 and N of the model to achieve optimum fitting accuracy. This will be described in the next section. The combined method presented in [6] does not work properly in this Li-ion battery experiment. Apparently, new parameters need to be used in the equations. Better results will probably be obtained when the model is adapted for this type of battery, but even then the disadvantage of using fixed parameter values will remain.

4.2.6 Summary

The proposed method of predicting the relaxation end voltage is to use the voltage-prediction model of Eq. (4.7), fit it to the first few minutes of the voltage-relaxation curve after a charge or discharge step, perform this fitting process in real-time (*i.e.* along the lines discussed in this section), and ignore the first part of the curve, *i.e.* the first 0.5 minute. The results presented in this section show that after 6 minutes of relaxation the EMF value can be predicted with an accuracy that allows the SoC value inferred from the EMF curve using the predicted EMF value to have an error of less than 1%.

A first advantage is fast and accurate assessment of SoC using the EMF method while the battery is not in equilibrium. Secondly, no pre-characterisation of a battery is needed to determine model parameters, as is the case with prior-art voltage-prediction systems. All the parameters in the proposed model described by Eq. (4.7) are learned on-line using the samples (t_i, V_{ti}) of the first few minutes of the relaxation process. The only parameters are t_1 (time at which the first voltage sample is taken) and N (number of samples used in the fitting process), which influence the fitting accuracy but have no direct link to the relaxation curve. Parameters in prior-art systems do have a direct link to the relaxation behaviour and must therefore be obtained via battery characterisation.

Thirdly, since the model parameters are calculated on-line, the dependencies of the voltage-relaxation on SoC, temperature, age, *etc.* are taken into account, which makes the model more widely applicable than prior-art voltage prediction models. Since the model is ideal for fitting relaxation curves with multiple time constants, it can also be used for battery chemistries other than Li-ion. A detailed study of the voltage-relaxation model as a function of battery aging will be given in chapter 6.

Fourthly, in addition to the relaxation end voltage the proposed voltage-prediction model can also predict the time it takes for the battery voltage to reach the predicted end value. This is not done in prior-art systems. This information can be used to tune the parameters t_1 and N, which describe which part of the relaxation curve is used for fitting, to achieve optimum accuracy.

Finally, by using the mean of the samples of the voltage-relaxation curve low-pass filtering is achieved, which makes the system less vulnerable to voltage spikes. This filtering can be done because the prediction is based on a few minutes of voltage-relaxation involving a large number of voltage points. The N data points remaining after low-pass filtering describe a voltage-relaxation curve that is considerably smoother than the curve described by all the measured points. The prior-art systems discussed in this chapter use instantaneous voltage samples, which means that when a voltage spike occurs at the sampling instant, wrong predictions are made. Low-pass filtering of voltage samples could of course be applied to those systems too.

4.3 Hysteresis

The charge/discharge EMF difference experimentally determined by means of the voltage-relaxation method described in section 4.1 will be investigated in this section. Fig. 4.16 presents the differences in charge/discharge EMFs obtained with the voltage-relaxation method at different temperatures. It follows from this figure that the maximum difference between the charge/discharge EMFs is 40 mV at around 5% SoC at 45°C. This means that when the EMF is used without taking the charge/discharge EMF difference into consideration, by *e.g.* modelling only the charge EMF curve, the SoC indication system based on the EMF will yield an SoC value of 5.1%, whereas the actual SoC value calculated on the basis of the discharge EMF is 6.1%. The inaccuracy will be 1% SoC. This effect will be more pronounced at low temperatures and in the flat region of the EMF-SoC curve, where even minor differences in the EMF will cause substantial errors in SoC.

Fig. 4.17 presents the differences in charge/discharge EMFs as a function of SoC at two C-rate currents. As can be seen, the difference in charge/discharge EMFs is consistently the same at SoC values higher than 4%. The small difference at low SoC values is explained by the different discharge C-rates used in the experiments. At higher C-rate currents the overpotential influence on the measured EMF curves becomes noticeable.

In order to further investigate the differences in charge/discharge EMFs three measurement artefacts that may have caused the difference were considered [8]. In the first place, due to the long rest periods chosen in the *voltage-relaxation* measurements, the self-discharge may have influenced the measured SoC-EMF values. Secondly, the battery voltage may not have reached the EMF equilibrium voltage by the end of the rest periods used in the *voltage-relaxation* experiment. Thirdly, in the *voltage-relaxation* method described in section 4.1 the different C-rates and time periods chosen for the charge/discharge steps in comparison with the

"Deep-Charge/Discharge" steps, *i.e.* 0.1 C-rate and 0.05 C-rate, respectively, may have influenced the measured SoC-EMF values.

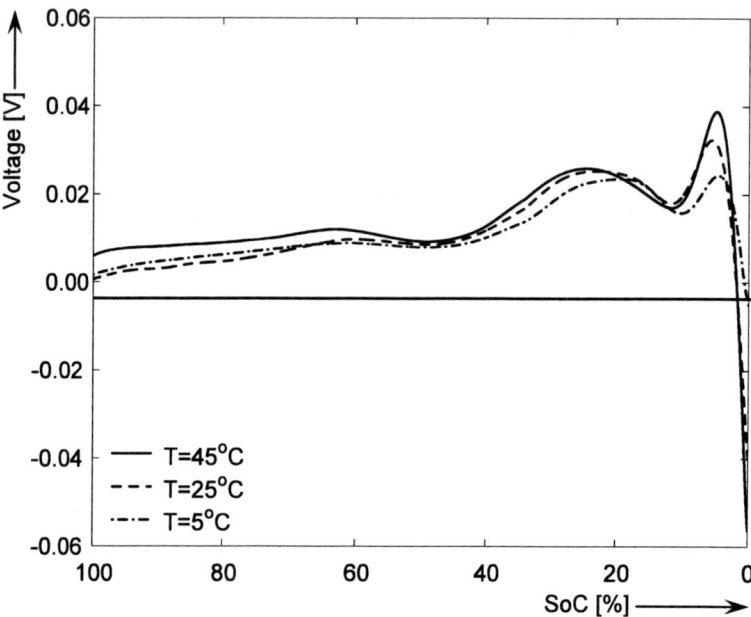

Fig. 4.16. The measured charge/discharge EMF difference as a function of SoC at three temperatures. The horizontal axis shows the SoC [%] normalised to maximum capacity.

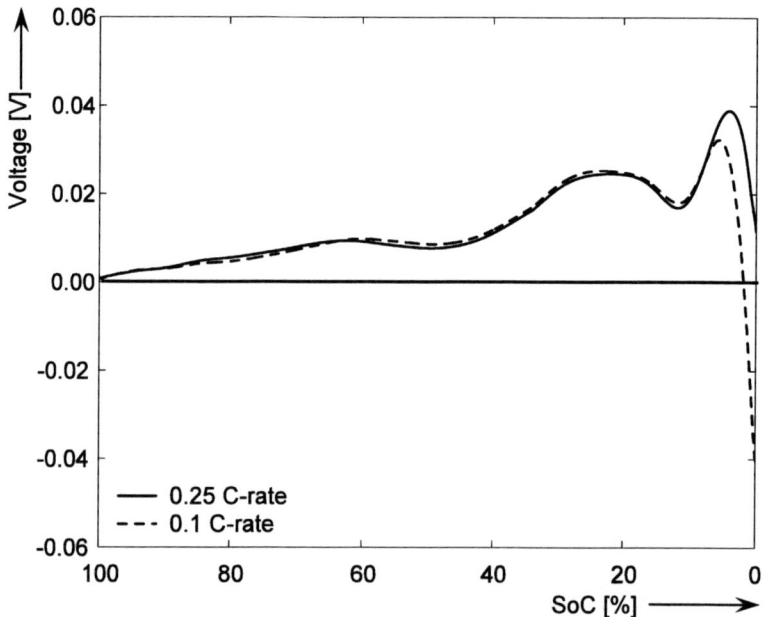

Fig. 4.17. The measured charge/discharge EMF difference as a function of SoC at two C-rate currents at 25°C.

The first measurement artefact was investigated by considering two uniform self-discharge values of 3% SoC/month and 6% SoC/month at 25°C [1]. To this end the SoC [%] values on the horizontal axis of the measured EMF curves were corrected by taking the two assumed self-discharge values into account. The new voltage-relaxation model presented in section 4.2 was used to investigate the second measurement artefact. And the third measurement artefact was investigated by performing battery initialisation measurements and new voltage-relaxation measurements using the Galvanostatic Intermittent Titration Technique (GITT).

The initialisation measurement was conducted as follows. First, a battery was fully charged using the usual Constant-Current-Constant-Voltage (CCCV) charging method at 0.5 C-rate. In the CV mode the voltage was kept constant at 4.2 V until the current reached a 0.05 C-rate value. At the end of the CV mode the SoC level was defined to be 100%. The charging step was followed by a rest period of about 4 hours, after which a discharge step was applied until the battery voltage reached 2.6 V at a constant 0.5 C-rate. The discharge step was followed by a rest period of 48 hours and a "Deep-Discharge" step at 0.001 C-rate until the battery voltage reached 2.6 V. After this "Deep-Discharge" step a rest period of 96 hours was applied. By the end of the rest period the battery voltage had reached the equilibrium voltage at a value of 3 V and the SoC level was defined to be 0%. The long rest periods were chosen to ensure that the equilibrium voltage, *i.e.* the EMF, would each time be reached. In addition, the maximum capacity was calculated by summating the total capacity obtained from the battery during the 0.5 C-rate and 0.001 C-rate discharging steps, respectively.

To obtain EMF-SoC values by means of GITT the battery was further charged at 0.1 C-rate in 25 steps. For each step in the experiment a limit of 4.3 V in voltage and 4% in SoC was assumed. Each charge step was followed by a rest period. The rest period was chosen as a function of SoC, *i.e.* at low SoC values long rest periods of 24 hours were chosen and at higher SoC values shorter rest periods of 12 hours. After charging, 25 discharge steps of 4% SoC each followed at a 0.1 C-rate and a voltage limit of 2.6 V. Each discharge step was followed by a rest period. The rest period was chosen as a function of SoC, *i.e.* at low SoC values long rest periods of 24 hours were chosen and at higher SoC values shorter rest periods of 12 hours where chosen. The low C-rate used in the experiments was chosen to obtain an equilibrium voltage faster. The advantage of the GITT method is that equal steps are considered during (dis)charging. Consequently, the build-up in time of the overpotential is also considered at different SoC levels. More information on the overpotential will be given in Chapter 5.

Fig. 4.18 illustrates what happens to the charge/discharge EMF difference when the battery self-discharge, the predicted EMF voltage and the GITT voltage-relaxation measurements are considered.

A first conclusion that can be drawn from Fig. 4.18 is that self-discharge is not the cause of the charge/discharge EMF difference. A second conclusion is that predicting the EMF voltage on the basis of the first part of the measured voltage-relaxation curve leads to smaller charge/discharge EMF differences at lower SoC values. This can be explained by the fact that at low SoC values (*i.e.* SoC values lower than 25%), rest periods of 24 hours were evidently not sufficient to obtain the equilibrium voltage.

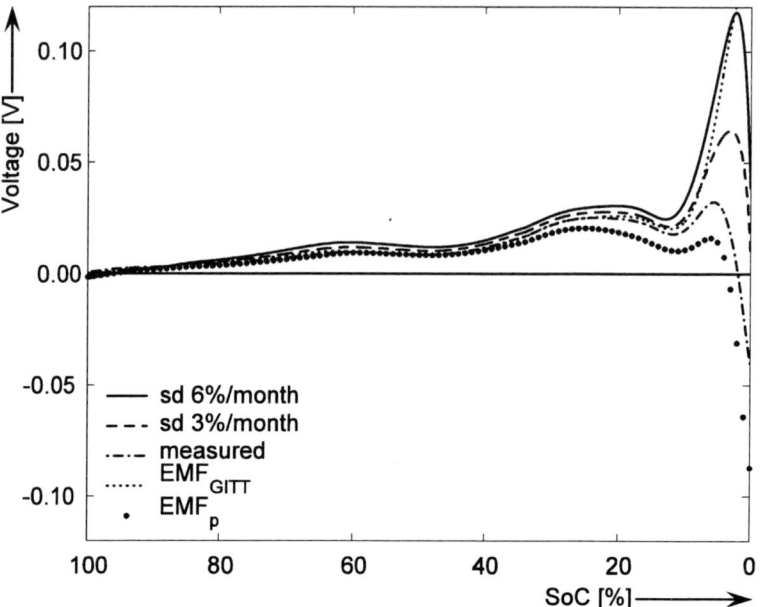

Fig. 4.18. The measured and calculated charge/discharge EMF difference at 25°C when the battery self-discharge (*sd*), the predicted EMF voltage (EMF_p) and the voltage-relaxation measurements by means of GITT (EMF_{GITT}) are considered.

The remaining small EMF difference may be explained by hysteresis between the charge and discharge EMF values. There are a few documented examples of hysteresis occurring in Li-ion battery systems [9]–[11]. They include the history-dependent equilibrium potential observed during the intercalation of lithium ions into carbon [10], [11]. In these articles it is assumed that the lithium atoms bind to hydrogen-terminated edges of hexagonal carbon fragments. This makes the capacity for the insertion of lithium strongly dependent on the hydrogen content of the carbon materials. If the inserted lithium binds to a carbon atom, which also binds to a hydrogen atom, a corresponding change will occur in the carbon-carbon bond, from $(sp)^2$ to $(sp)^3$. This bonding change in the host will lead to pronounced hysteresis during lithium insertion [10], [11].

We do not claim to understand this complex process. The hysteresis may also be introduced by the *LiCoO₂* electrode. It may be tentatively concluded that a possible cause of the hysteresis could be the occurrence of phase transitions (*ph*). For further information on electrochemical hysteresis the reader is referred to [10]–[12].

4.4 Electro-Motive Force modelling

A model for the EMF-SoC relationship will be presented in this section. This model can be used to calculate the SoC at a certain EMF and temperature. The EMF curves obtained by means of GITT as discussed in the previous section are approximated using a mathematical function in which the EMF of a Li-ion battery with intercalated electrodes is modelled as the difference in equilibrium potentials between the positive and negative electrodes, according to

$$EMF = E_{eq}^+ - E_{eq}^- \qquad (4.12)$$

where the equilibrium potential of the positive electrode (E_{eq}^+) is given by

$$E_{eq}^+ = E_0^+ - \frac{RT}{F}\left(\log\left(\frac{x_{Li}}{1-x_{Li}}\right) + U_j^+ x_{Li} - \zeta_j^+\right) \qquad (4.13)$$

$$\zeta_2^+ = (U_2^+ - U_1^+)x_{ph} + \zeta_1^+, \quad j = \begin{cases} 1, & x_{ph} \leq x_{Li} \leq 1 \\ 2, & 1/2 \leq x_{Li} \leq x_{ph} \end{cases} \qquad (4.14)$$

in which E_0^+ is the standard redox potential of the $LiCoO_2$ electrode in [V], U_j^+ denotes the dimensionless interaction energy coefficient in the $LiCoO_2$ electrode, ζ_j^+ is a dimensionless constant, x_{Li} the molar fraction of Li^+ ions inside the positive electrode corresponding to the SoC of the $LiCoO_2$ electrode, R is the gas constant [J (mol K)$^{-1}$], F the Faraday constant [C mol^{-1}] and T is the (ambient) temperature in [K]. In Eq. (4.14) a phase transition (ph) occurs at $x_{Li}=x_{ph}$ that results in a curvature change. According to recent literature on Li-ion batteries with an $LiCoO_2$ electrode (see e.g. [1], [13]) the main phase-transition point lies near $x_{ph} \approx 0.75$. A phase transition is observed as a change in the slope of the equilibrium potential as a function of x_{Li}. This change in the slope is realized in the present physical model by the change in interaction energy between the intercalated Li^+ ions from a value U_1^+ in phase 1 to a value U_2^+ in phase 2 [1]. The values of the dimensionless constants ζ_1^+ and ζ_2^+ in phases 1 and 2 are chosen so that a continuous transition in equilibrium potential between phases 1 and 2 is obtained (see Eq. (4.14)).

The negative electrode is modelled in the same way

$$E_{eq}^- = E_0^- - \frac{RT}{F}\left(\log\left(\frac{z_{Li}}{1-z_{Li}}\right) + U_j^- z_{Li} - \zeta_j^-\right) \qquad (4.15)$$

$$\zeta_2^- = (U_2^- - U_1^-)z_{ph} + \zeta_1^-, \quad j = \begin{cases} 1, & 0 \leq z_{Li} \leq z_{ph} \\ 2, & z_{ph} \leq z_{Li} \leq 1 \end{cases} \qquad (4.16)$$

where E_0^- is the standard redox potential of the LiC_6 electrode in [V], U_j^- denotes the dimensionless interaction energy coefficient in the LiC_6 electrode, ζ_j^- is a dimensionless constant and z_{Li} is the molar fraction of the Li^+ ions inside the negative electrode. A phase transition occurring around $z_{ph} \approx 0.25$ is modelled in the negative electrode. As in the case of the positive electrode, the phase transition is observed as a change in the slope of the equilibrium potential as a function of z_{Li}. This change in slope of the equilibrium potential is in the present physical model realized by a change in interaction energy between the intercalated Li^+ ions from a value U_1^- in phase 1 to a value U_2^- in phase 2 [1]. The values of the dimensionless constants ζ_1^- and ζ_2^- in phases 1 and 2, respectively, are chosen so that a continuous transition is achieved from the equilibrium potential in phase 1 to that in phase 2 (see Eq. (4.16)). Under normal operating conditions x_{Li} will cycle between 0.5 and 1 and z_{Li} between 0 and 1 [1].

Besides parameters characterising the electrode electrochemistry, parameters relating to the battery's design are also needed to model the EMF-SoC relationship in a practical SoC indication system. As schematically indicated in Fig. 4.19, a fresh Li-ion battery can be characterised by the maximum capacity of the positive electrode (Q_{max}^+), the maximum capacity of the negative electrode (Q_{max}^-), the number of electrochemically active Li^+ ions inside a fresh battery (Q_{max}) and the number of Li^+ ions inside the negative electrode in a "fully" discharged (under standard operating conditions) battery (Q_0^-). Finally, Q_z^- denotes the charge stored in the negative electrode at a given SoC.

In a new battery the number of electrochemically active Li^+ ions inside the battery Q_{max} will correspond to the maximum capacity of the positive electrode Q_{max}^+. It should be noted that because x_{Li} cycles between 0.5 and 1 only half of the maximum capacity of the positive electrode Q_{max}^+ is cyclable. The number of cyclable electrochemically active Li^+ ions therefore equals $Q_{max} - Q_{max}^+/2$. During the first activation cycles a portion of the Li^+ ions will remain in the negative electrode (represented by Q_0^- in Fig. 4.19) and a portion will be consumed in the Solid Electrolyte Interface (SEI), in an irreversible process. The Q_0^- capacity can be explained by the Nernstian decrease in the LiC_6 electrode equilibrium voltage when its SoC goes to zero. The SEI suppresses the decomposition of the electrolyte at electrode surface. For simplicity the SEI has not been illustrated in Fig. 4.19 and is not considered in the present physical model. As a result, the value of the maximum capacity of the positive electrode Q_{max}^+ will be larger than the capacity corresponding to the number of electrochemically active Li^+ ions inside the battery (see Fig. 4.19).

Methods for measuring and modelling a battery's Electro-Motive Force

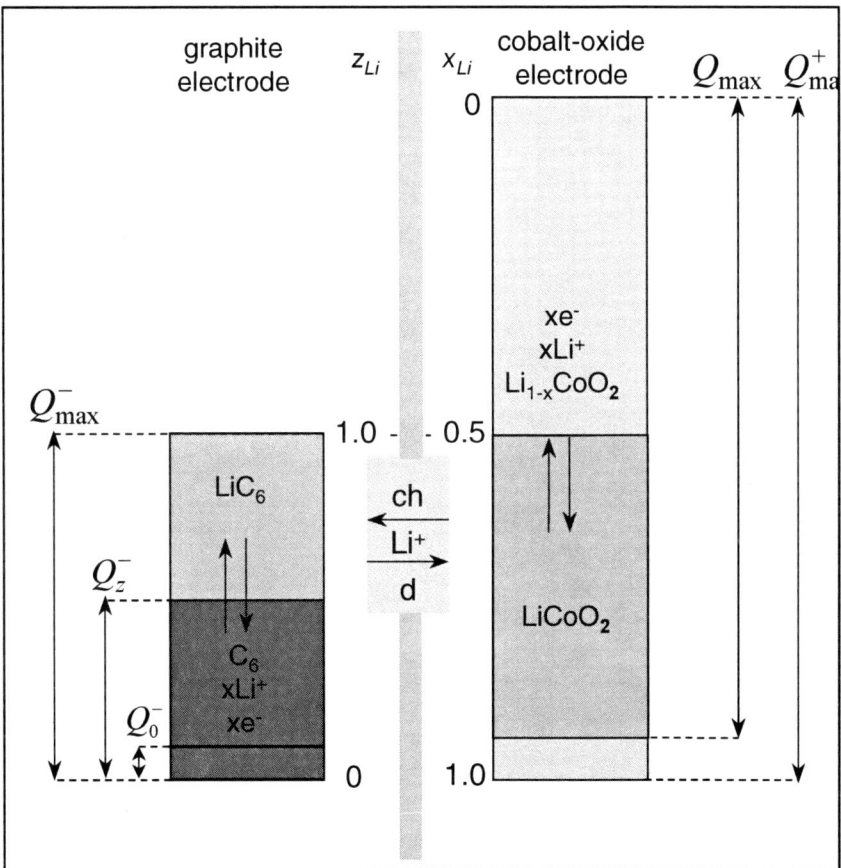

Fig. 4.19. Schematic representation of the EMF-SoC relationship parameters relating to battery design.

Given the parameters shown in Fig. 4.19 and the experimentally determined SoC values, x_{Li} and z_{Li} can be inferred from

$$z_{Li} = \frac{Q_z^-}{Q_{max}^-} \tag{4.17}$$

$$Q_z^- = Q_0^- + \frac{SoC}{100}\left(Q_{max}^- - Q_0^- - \frac{Q_{max}^+}{2}\right) \tag{4.18}$$

$$x_{Li} = \frac{Q_{max} - Q_z^-}{Q_{max}^+} = \frac{Q_{max} - z_{Li}Q_{max}^-}{Q_{max}^+} \tag{4.19}$$

Equations (4.17)–(4.19) can be understood as follows. It follows from Fig. 4.19 that the Li^+ ions move from the cobalt-oxide electrode to the graphite electrode during charging. At the end of charging the battery SoC is defined to be 100%, i.e. Q_z^- is equal to Q_{max}^- and Q_z^- equals $Q_{max}^- - Q_{max}^+/2$ (Eq. (4.17)). It follows from Eqs. (4.17) and (4.19) that at 100% SoC $x_{Li} = [Q_{max} - (Q_{max} - Q_{max}^+/2)]/Q_{max}^+ = 0.5$ and $z_{Li}=1$. During discharge the Li^+ ions move from the graphite electrode to the cobalt-oxide electrode. At the end of discharging under standard operating conditions, $Q_z^- = Q_0^-$. As can be inferred from Eqs. (4.17) and (4.19), the $x_{Li} = (Q_{max} - Q_0^-)/Q_{max}^+$ will be a somewhat smaller than 1 and $z_{Li} = Q_0^-/Q_{max}^-$ will be a little larger than 0.

E_{eq}^+ and E_{eq}^- can be inferred from Eqs. (4.13), (4.14) and (4.15), (4.16), respectively. However, the EMF measurements showed that the phase transition does not take place instantaneously, but in a certain interval around the phase transition points. In order to achieve a smooth phase transition the following approximation was considered

$$E_{eq}^+ = E_0^+ - \frac{RT}{F}\left\{\log\left(\frac{x_{Li}}{1-x_{Li}}\right) + \Phi\left(\frac{x_{Li}-x_{ph}}{\sigma_x}\right)(U_1^+ x_{Li} - \varsigma_1^+) + \left[1-\Phi\left(\frac{x_{Li}-x_{ph}}{\sigma_x}\right)\right](U_2^+ x_{Li} - \varsigma_2^+)\right\}$$

(4.20)

$$E_{eq}^- = E_0^- - \frac{RT}{F}\left\{\log\left(\frac{z_{Li}}{1-z_{Li}}\right) + \Phi\left(\frac{z_{Li}-z_{ph}}{\sigma_z}\right)(U_1^- z_{Li} - \varsigma_1^-) + \left[1-\Phi\left(\frac{z_{Li}-z_{ph}}{\sigma_z}\right)\right](U_2^- * z_{Li} - \varsigma_2^-)\right\}$$

(4.21)

where Φ denotes a standard normal cumulative distribution function and the parameters σ_x and σ_z determine the smoothness of the phase transitions in the positive and negative electrode, respectively.

In order to include the influence of temperature in the EMF-SoC relationship a linear dependence of each model parameter (*par*) was assumed according to

$$par(T) = par(T_{ref}) + (T - T_{ref})\Delta par \quad (4.22)$$

where T_{ref} is the reference temperature (in this case 25°C) and T is the ambient temperature. Δpar is each parameter's sensitivity to temperature.

Different values for the model parameters can be used for the charge- and discharge-EMF in order to deal with the hysteresis effect. When another type of Li-ion battery with a different EMF-SoC curve chemistry is used the model can be adapted by fitting, leading to new parameter values. This model is consequently not limited to the present type of Li-ion battery. Taking into account hysteresis and temperature, the presented method is assumed to be the best solution for practical EMF implementation.

Methods for measuring and modelling a battery's Electro-Motive Force

The results of the mathematical EMF implementation will be presented in the remainder of this chapter. The EMF equations described in this section need to be fitted to a measured EMF curve. Fig. 4.20 shows that the modelled EMF curve used in the system shows a good fit with the measured curve obtained with the Maccor battery tester via voltage relaxation according to the protocol discussed in section 4.1 at all temperatures.

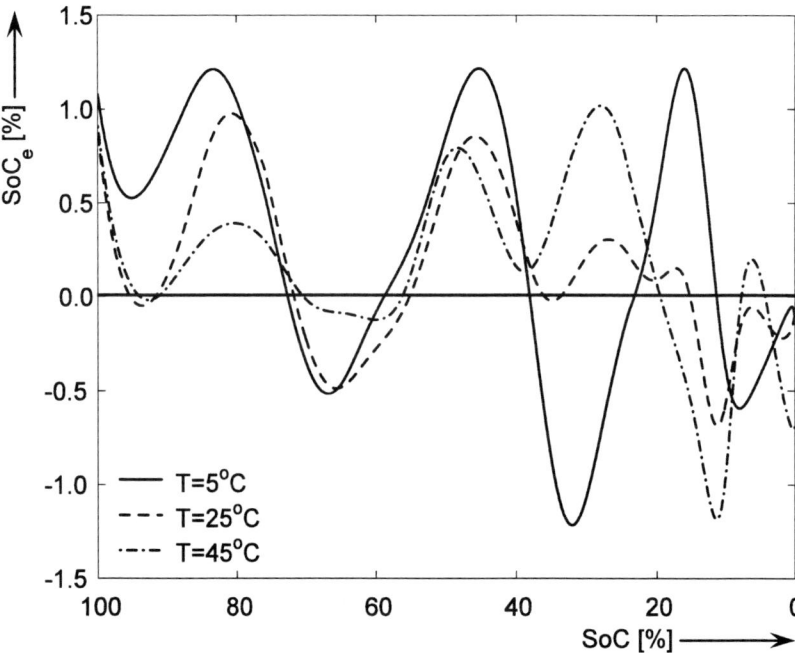

Fig. 4.20. Accuracy of SoC indication using the EMF obtained via the voltage-relaxation curve versus the fitted EMF curve at 5, 25 and 45°C. The horizontal axis shows the SoC [%] normalised to maximum capacity.

It can be concluded that the maximum error in SoC, SoC_e, is obtained at 5°C and around 16% SoC. This 1.2% SoC error corresponds to a 13.2 mAh capacity value, which can still be removed from the battery. As can be calculated using

$$t_r[\min] = \frac{SoC_e[\%]}{100} \frac{1}{C_d} 60 \qquad (4.23)$$

for a new battery at a 0.5 C-rate mean discharge current C_d this implies around 1.4 minutes of remaining run-time t_r.

However, the SoC error will usually be smaller than 0.9% so it can be concluded that the error obtained in EMF implementation using the model presented in this section will generally provide enough accuracy to achieve a final accuracy of within 1.1 minutes or better in remaining run-time indication.

The EMF model parameters values used in the simulations illustrated in Fig. 4.20 are summarised in Table 4.3. Column one gives the symbol of the EMF model parameter. The values and units of the EMF model parameters are given in columns two and three, respectively. It should be noted that the ζ_2^+ and ζ_2^- parameters were derived from Eqs. (4.14) and (4.16), respectively.

Table 4.3. The battery EMF model parameter values.

Parameter	Value	Unit
U_1^+	9.68 10^2	[1]
ΔU_1^+	−1.89	[T^{-1}]$^+$
U_2^+	1.63 10^4	[1]
ΔU_2^+	−5.14 10^1	[T^{-1}]$^+$
U_1^-	−6.92 10^3	[1]
ΔU_1^-	2.18 10^1	[T^{-1}]$^+$
U_2^-	−6.89 10^3	[1]
ΔU_2^-	2.16 10^1	[T^{-1}]$^+$
ζ_1^+ †	0.00	[1]
$\Delta\zeta_1^+$	2.22 10^{-4}	[T^{-1}]$^+$
ζ_1^-	−2.14 10^{-5}	[1]
$\Delta\zeta_1^-$	2.04 10^{-4}	[T^{-1}]$^+$
x_{ph}	8.44 10^{-1}	[1]
Δx_{ph}	1.58 10^{-6}	[T^{-1}]$^+$
z_{ph}	4.44 10^{-1}	[1]
Δz_{ph}	1.37 10^{-4}	[T^{-1}]$^+$
σ_{x1}	1.72 10^{-2}	[1]
$\Delta\sigma_{x1}$	−2.24 10^{-6}	[T^{-1}]$^+$
σ_{z1}	1.22 10^{-1}	[1]
$\Delta\sigma_{z1}$	8.98 10^{-4}	[T^{-1}]$^+$
$E_0^+ - E_0^-$ *	1.12 10^1	[V]
$\Delta(E_0^+ - E_0^-)$ *	2.52 10^{-2}	[V T^{-1}]$^+$
Q_{max}^+ #	1.00	[1]
Q_{max}^-	4.23 10^{-1}	[1]
Q_{max}	8.11 10^{-1}	[1]
Q_0^-	9.63 10^{-4}	[1]
ΔQ_0^-	4.34 10^{-5}	[T^{-1}]$^+$

$^+$ See equation 4.22.
† Parameter set to zero by identification condition.

* Only the $E_0^+ - E_0^-$ difference is identifiable in the battery EMF obtained by means of the voltage-relaxation measurement method described in this chapter. The E_0^+ and E_0^- can be separately inferred from voltage-relaxation measurements conducted for individual electrodes. Such an experiment is not considered in this book, so the $E_0^+ - E_0^-$ difference was assumed to be an estimable parameter.

\# The Q_{max}^+ parameter was not inferred from the battery EMF obtained by means of the voltage-relaxation measurement method described in this chapter. The Q_{max}^+ was not estimated but normalised to be one. The other Q-related parameters were inferred as fractions from the Q_{max}^+ value. Those parameters are consequently all dimensionless.

It should be noted that the scope of the research discussed in this book was to achieve a high level of accuracy in SoC determination (see chapter 1). The EMF model parameters values presented in Table 4.3 were determined during the fitting process with this in mind. For this reason these parameters are in some cases out of their physical range.

4.5 Conclusions

Methods for measuring and modelling the EMF function of Li-ion batteries have been presented in this chapter. The EMF function is used as input for a State-of-Charge algorithm that calculates the SoC in percentages as well as the remaining run-time for portable applications.

Two EMF determination methods have been compared. Of the two, the voltage-relaxation method is recommended for EMF determination. A new voltage-relaxation model that speeds up the SoC indication on the basis of EMF has been developed. It has been found that after current interruption the first 0.5 minute must be ignored to obtain a high level of voltage prediction accuracy. From the low calculated SoC error, *i.e.* always less than 1% when the voltage prediction starts after a rest period of five minutes, it can be concluded that the developed voltage-relaxation model is the best solution for obtaining an accurate EMF value in a short time and under a wide range of conditions.

The effects of the temperature and hysteresis on the EMF-SoC curve and the influences on the SoC indication accuracy have been discussed. A mathematical model has been presented for the SoC-EMF relationship that also takes the effect of temperature into consideration. Good EMF fitting results have been obtained. These results yield an EMF prediction that is always better than 1.2% SoC and a remaining run-time prediction that is always better than 1.4 minutes.

4.6 References

[1] H.J. Bergveld, W.S. Kruijt, P.H.L. Notten, Battery Management Systems, Design by Modelling, Philips Research Book Series, **1**, Kluwer Academic Publishers, Boston (2002)

[2] F.A.C.M. Schoofs, W.S. Kruijt, R.E.F. Einerhand, S.A.C. Hanneman, H.J. Bergveld, Method of and Device for Determining the Charge Condition of a Battery, US Patent 6,420,851, filed March 29 (2000)

[3] H.J. Bergveld, H. Feil, J.R.G.C.M. Van Beek, Method of predicting the state of charge as well as the use time left of a rechargeable battery, US Patent 6,515,453, filed November 30 (2000)

[4] P.P.L. Regtien, F. van der Heijden, M.J. Korsten and W. Olthuis, Measurement Science for Engineers, Kogan Page Science Publisher, London, (2004)

[5] J.H. Aylor, A. Thieme, B.W. Johnson, A Battery State-of-Charge Indicator for Electric Wheelchairs, IEEE Trans. on industrial electronics, **39**, 398–409 (1992)

[6] S. Hoenig, H. Singh, T.G. Palanisamy, Method for determining state of charge of a battery by measuring its open circuit voltage, US Patent 6,366,054, filed May 2 (2001)

[7] H.J. Bergveld, V. Pop and P.H.L. Notten, Apparatus and method for determination of the state-of-charge of a battery when the battery is not in equilibrium, Patent PH006795EP1, filed October 30 (2006)

[8] V. Pop, H.J. Bergveld, J.H.G. Op het Veld, D. Danilov, P.P.L. Regtien, P.H.L. Notten, Modelling battery behaviour for accurate State-of-Charge indication, J. Electrochem. Soc., **153**, A2013–A2022 (2006)

[9] H. Wonchull, H. Mitsuhiro, K. Tetsuichi, Hysteresis on the electrochemical lithium insertion and extraction of hexagonal tungsten trioxide. Influence of residual ammonium, Solid State Ionics Journal, **128**, 25–32 (2000)

[10] T. Zheng, W.R. McKinnon, J.R. Dahn, Hysteresis During Lithium Insertion in Hydrogen-Containing Carbons, J. Power Sources, **143**, 2137–2145 (1996)

[11] T. Zheng, J.R. Dahn, Hysteresis Observed in Quasi-open Circuit Voltage Measurements of Lithium Insertion in Hydrogen-containing Carbons, J. Power Sources, **68**, 201–203 (1997)

[12] V. Srinivasan, J. W. Weidner, J. Newman, Hysteresis during Cycling of Nickel Hydroxide Active Material, J. of Electrochem. Soc., **148**, 969–980 (2001)

[13] G.A. Nazri, G. Pistola, Science and Technology of Lithium Batteries (2002)

Chapter 5
Methods for measuring and modelling a battery's overpotential

During the charge and discharge states a battery's voltage does not equal its Electro-Motive Force (EMF). The difference between the EMF and the voltage during current-flowing conditions is defined as the overpotential. Measurements, modelling and simulation results obtained for battery overpotentials are presented in this chapter. Overpotential measurements involving partial and full charge/ discharge steps are presented in section 5.1. Overpotential symmetry, a phenomenon discovered during the analysis of the measurement results, is also presented in this section. A new overpotential model that includes C-rate current dependency and the simulation results obtained with this model are presented in section 5.2. These results are also compared with the measurement results. Finally section 5.3 presents concluding remarks.

5.1 Overpotential measurements

A battery's overpotential is defined as the difference between the battery's EMF and its charge/discharge voltage [1]–[3]. Due to this overpotential, a battery's voltage during the (dis)charge state is (lower) higher than the EMF voltage. The value of a battery's overpotential depends on the charge/discharge current value, charge/discharge period, SoC, temperature, battery chemistry and aging. Two methods for measuring the battery overpotential will be presented in this section. The measurements presented in this chapter were carried out using fresh Sony US18500G3 Li-ion batteries.

5.1.1 Overpotential measurements involving partial charge/discharge steps

As the overpotential represents the difference between a battery's EMF and its charge/discharge voltage, a charge/discharge EMF was first experimentally determined by means of the Galvanostatic Intermittent Titration Technique (GITT) described in section 4.3. During this measurement the battery was charged/ discharged in equal steps of 4% SoC at 0.1 C-rate current and 25°C. The 4% SoC was inferred from the charge/discharge capacity obtained by means of Coulomb counting [1]–[7] and from the battery's maximum capacity. Each charge/discharge step was followed by rest periods of 24 hours at low SoC values and 12 hours at high SoC values. It was assumed that the battery voltage had reached the equilibrium voltage value, *i.e.* the EMF, by the end of the rest periods. In order to obtain information on the battery's overpotential the difference between the battery's EMF and the voltage last measured after a charge/discharge step was considered. The charge overpotential consequently had a negative value and the discharge overpotential a positive value.

Fig. 5.1 shows a battery overpotential calculated as the difference between the EMF and the voltage last measured after a charge (V_{ch}) step of 25 minutes at 28% SoC, 0.1 C-rate current and 25°C. It can be concluded from Fig. 5.1 that the charge battery overpotential η_{ch} in this example is −37 mV.

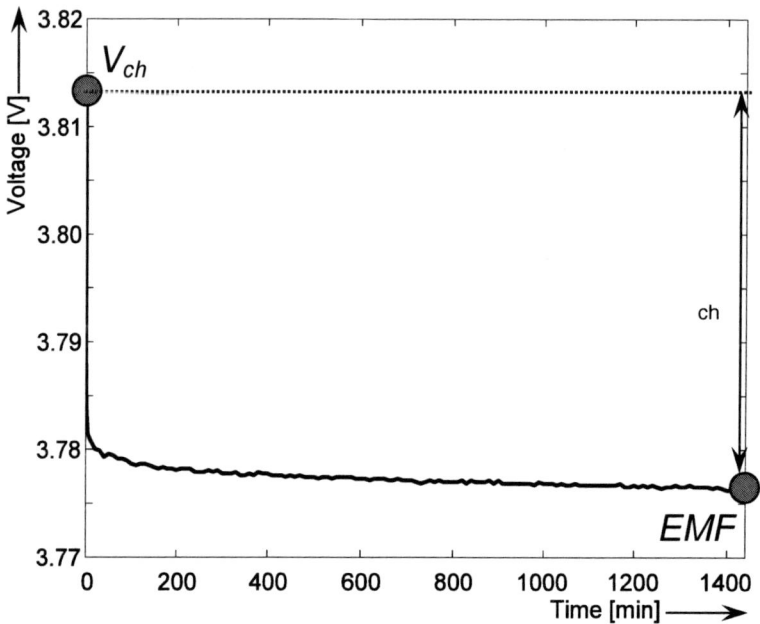

Fig. 5.1. Battery charge overpotential (η_{ch}) calculated as the difference between the EMF and the voltage last measured after a charge (V_{ch}) step of 25 minutes at 28% SoC, 0.1 C-rate current and 25°C. The horizontal axis shows the time in [min.].

An important advantage of using GITT voltage-relaxation measurements for overpotential calculations is that interpolation calculations are not needed for the charge/discharge battery overpotential comparison. The difference between the EMF and the voltage last measured after a charge/discharge step is calculated at each 4% SoC. The inaccuracy that could result from the measurement interpolation is consequently eliminated from the charge/discharge battery overpotentials comparison. Fig. 5.2 shows the battery overpotential calculated after 25-minute charge and discharge steps at 0.1 C-rate current, 4% SoC and 25°C. It can be concluded from Fig. 5.2 that the maximum overpotential after the discharge step is 112 mV.

The GITT voltage-relaxation measurements showed that, especially at low SoC values, a rest period of 24 hours was not always sufficient to ensure equilibrium voltage. In order to investigate the influence of a not-fully-relaxed voltage in the battery overpotential calculation, the battery voltage was further predicted by using the new voltage-relaxation model developed in section 4.2.3. The voltage values measured during the rest periods, *i.e.* 24 hours at low SoC values and 12 hours at high SoC values, were used as input for the voltage-relaxation model.

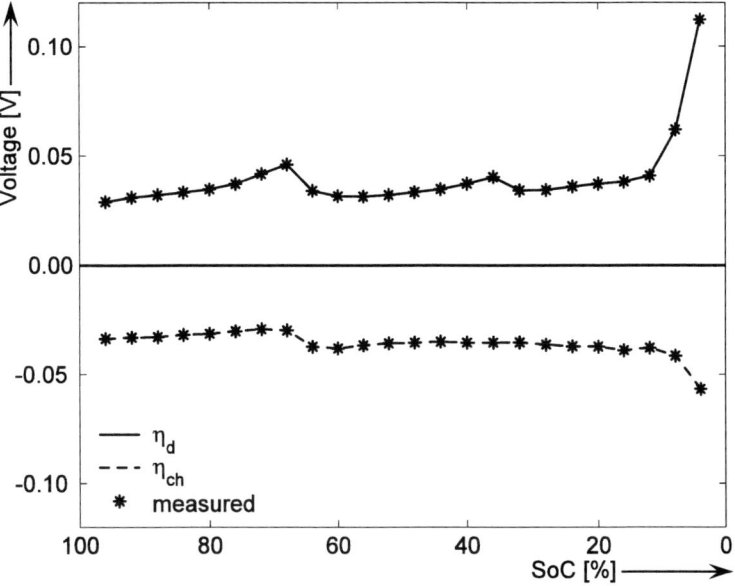

Fig. 5.2. The charge (η_{ch}) and discharge (η_d) battery overpotentials calculated as the difference between the EMF voltage and the voltage last measured after a charge/discharge step at each 4% SoC (*). The horizontal axis shows the SoC [%] normalised to maximum capacity.

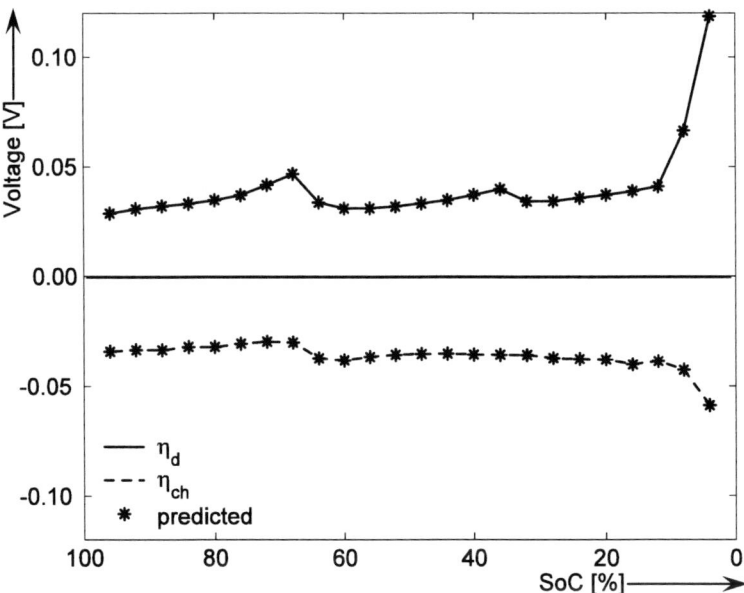

Fig. 5.3. Same as Fig. 5.2 when the predicted battery voltage is considered.

In this example a period of 96 hours was considered for the voltage prediction (see section 4.2.3). Fig. 5.3 shows the battery charge/discharge overpotentials calculated as the difference between the predicted battery voltage after 96 hours and the voltage last measured after a charge/discharge step.

It follows from Fig. 5.3 that a maximum discharge overpotential of 118 mV is obtained at 4% SoC. It can be concluded from Figs. 5.2 and 5.3 that the overpotential determined as the difference between the EMF and the V_{ch} and the charge overpotential determined as the difference between the predicted voltage (V_p) and the V_{ch} are consistently the same at all the SoC values. Similarly the discharge overpotential determined as the difference between the EMF and the voltage last measured after a discharge (V_d) step and the discharge overpotential determined as the difference between the V_p and the V_d are consistently the same at SoC values higher than 8%. Only a small voltage difference was observed at low SoC values. It can be concluded that rest periods of 24 hours at low SoC values and 12 hours at high SoC values are generally sufficient to ensure an accurate battery overpotential.

It can be concluded from the partial (dis)charge step measurements shown in Figs. 5.2 and 5.3 that the battery overpotential increases at low SoC values and remains almost constant at SoC values higher than 12%. Another important conclusion relates to the shape of the battery (dis)charge overpotentials. It follows from Figs. 5.2 and 5.3 that the calculated mean charge and discharge battery overpotentials are symmetrical between 20 and 80% SoC. More information on the overpotential symmetry will be given later in this chapter.

Partial (dis)charge battery overpotential measurements as a function of temperature will be presented below. To obtain information on this temperature dependence the previously described measurements at 25°C were repeated at 5 and 45°C. Fig. 5.4 shows the battery overpotential calculated after charge and discharge intervals of 4% SoC, 0.1 C-rate current and as a function of three temperatures. As can be seen in this figure, a maximum overpotential of 200 mV was obtained after a discharge step at 7% SoC and 5°C.

In order to investigate the influence of a not-fully-relaxed voltage in the battery overpotential calculation, the battery voltage at each temperature was further predicted by using the new voltage-relaxation model developed in section 4.2.3. The voltage values measured at 5, 25 and 45°C during the rest periods, *i.e.* 24 hours at low SoC values and 12 hours at high SoC values, were used as input for the voltage-relaxation model. In this example a rest period of 96 hours was considered for the voltage prediction. Fig. 5.5 shows the battery (dis)charge overpotentials calculated as the difference between the predicted battery voltage after 96 hours and the voltage last measured after a (dis)charge step at 5, 25 and 45°C.

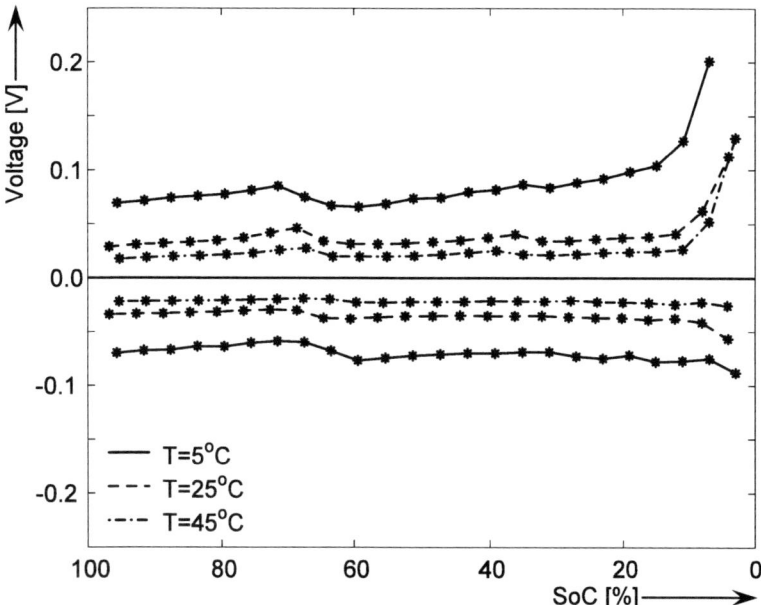

Fig. 5.4. The charge and discharge battery overpotentials calculated as the difference between the EMF voltage and the voltage last measured after a (dis)charge step at each 4% SoC (*) as a function of SoC at three temperatures.

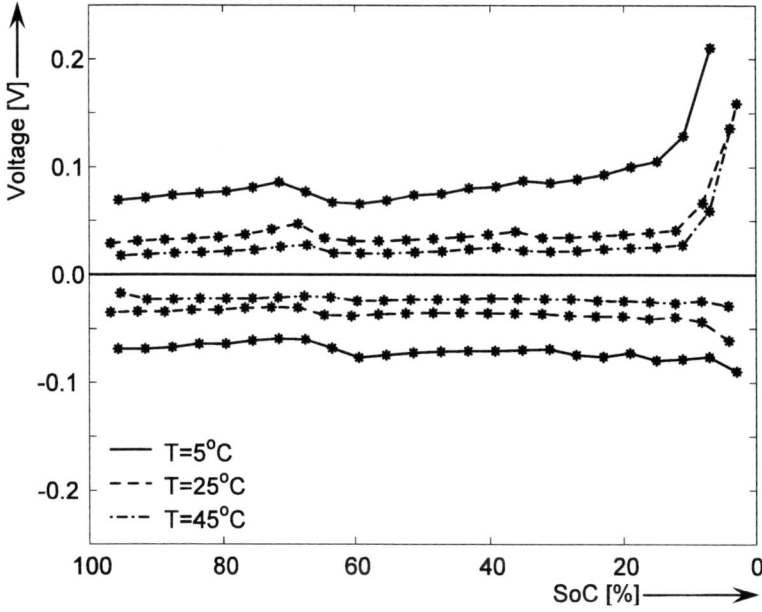

Fig. 5.5. Same as Fig. 5.4 when the predicted battery voltage is considered.

It follows from Figs. 5.4 and 5.5 that the overpotential increases at low SoC values and temperatures. In this case a maximum overpotential of 200 mV was obtained after a discharge step at 5°C. It can be concluded that the (dis)charge overpotentials determined by means of $V_d(V_{ch})$ are consistently the same at all the SoC values. Only a small voltage difference was observed at low SoC values for the discharge overpotential. This means that rest periods of 24 hours at low SoC values and 12 hours at high SoC values are generally sufficient to ensure an accurate battery overpotential.

It can be concluded from the partial (dis)charge step measurements and Figs. 5.2, 5.3, 5.4 and 5.5 that the battery overpotential increases at low SoC values, and remains almost constant at SoC values higher than 12%. The main drawback of the partial (dis)charge-step measurements is that complete information on the build-up of the overpotential in time cannot be obtained by using (dis)charge periods of 25 minutes.

5.1.2 Overpotential measurements involving full (dis)charge steps

In order to obtain information on the (dis)charge overpotential build-up in time, the difference between the (dis)charge EMF voltage experimentally determined using GITT and the full (dis)charge voltage curves was considered. The EMF was experimentally determined after a (dis)charge step by using GITT as described in section 4.3. Several full (dis)charge steps were carried out at 25°C. During these steps the battery was each time fully charged using the normal Constant-Current-Constant-Voltage (CCCV) charging method at 0.5 C-rate current. In the CV mode the voltage was kept constant at 4.2 V until the current reached a 0.05 C-rate value. The SoC level was defined to be 100% at the end of the CV mode. The same 0.5 C-rate charge current was used in all the experiments to ensure that the same 100% SoC level would each time be reached. Each charging step was followed by a rest period of 4 hours. After this rest a discharge step was applied until the battery voltage reached 3 V at constant C-rate currents of 0.1, 0.25, 0.5, and 1 C-rate current, respectively. Each discharge step was followed by a rest period of 12 hours and by a "Deep-Discharge" step at 0.01 C-rate current. After this "Deep-Discharge" step a rest period of 12 hours was applied. The "Deep-Discharge" step was performed to ensure that a new charge cycle would each time start at the same voltage and SoC value. The long rest steps were used to ensure that a new cycle would each time start at equilibrium voltage. In this way the effects of different initial SoC and voltage levels and of a not-fully-relaxed voltage were eliminated from the battery overpotential determination.

Fig. 5.6 shows the overpotential curves obtained for discharge at 25°C. All the overpotential curves were calculated as the difference, at the same SoC value, between the discharge EMF experimentally determined by means of GITT at 25°C and the discharge voltage curves measured at 25°C using the constant C-rate current method.

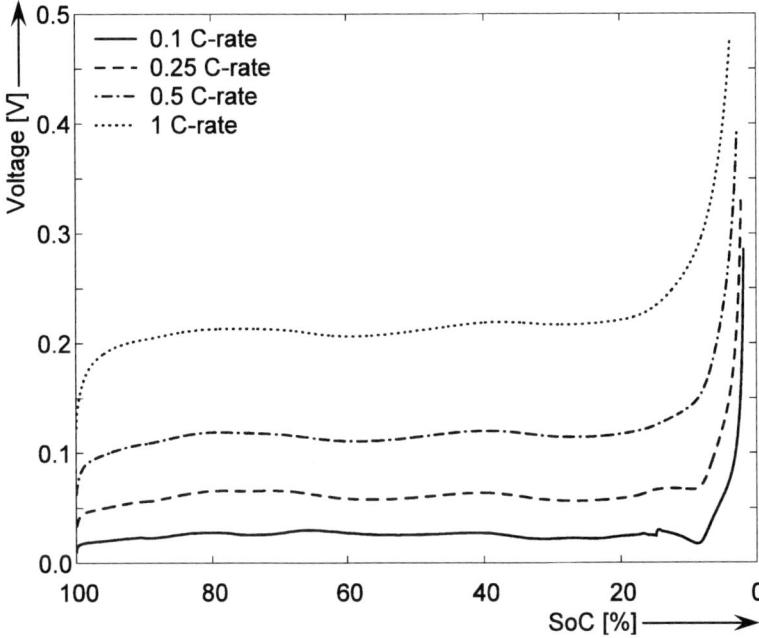

Fig. 5.6. Overpotential curves obtained at different C-rate currents at 25°C during discharge as a function of SoC.

It can be concluded from Fig. 5.6 that the overpotential increases more or less proportionally with the C-rate. Between 20 and 100% SoC the overpotential is moreover almost independent of the SoC, but it rises sharply at lower SoC values; a maximum overpotential of 475 mV was obtained after a discharge step at 1 C-rate current and 3.8% SoC.

In order to obtain information on the battery overpotential dependence on temperature, the previously described measurements were repeated at 5 and 45°C. Fig. 5.7 presents the battery overpotential obtained with the above-described measurement method at three different temperatures during discharging cycles at 0.25 C-rate. It follows from Fig. 5.7 that the overpotential increases for low temperatures. As an example when using the overpotential without taking into consideration the temperature effect by modelling only the overpotential curve at 25°C, an overpotential value of 330 mV at 5°C will be calculated, when actually the overpotential value is 559 mV. The inaccuracy, calculated as the difference between the true overpotential value measured at 5°C and the overpotential value measured at 25°C, will be 229 mV [9].

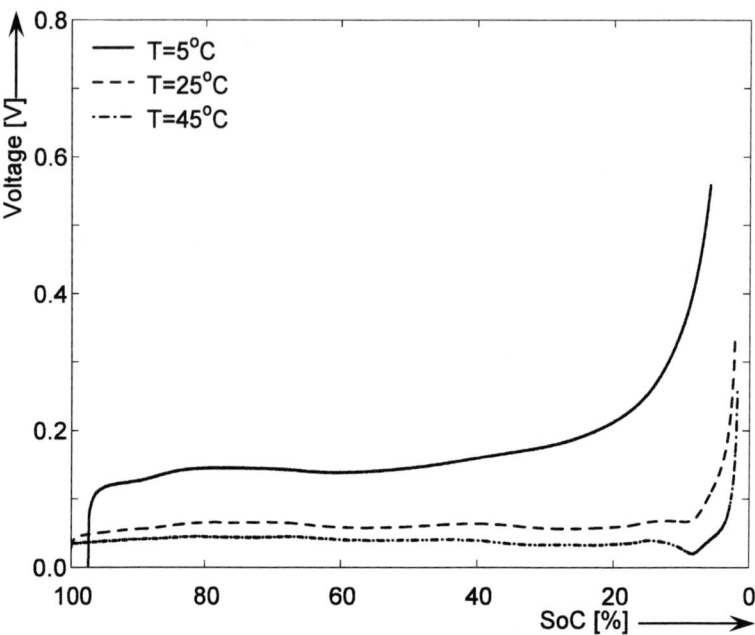

Fig. 5.7. Overpotential curves for different temperatures obtained at 0.25 C-rate during discharge. The horizontal axis shows the SoC [%] normalized to the maximum capacity.

As shown in the previous section, the overpotential measured by means of partial (dis)charge steps was symmetrical. This overpotential symmetry in the full (dis)charge step measurement method will be further discussed below. Fig. 5.8 shows the overpotential curves obtained at different discharge currents at 25°C and the overpotential curve obtained at 1 C-rate charge current at 25°C. All the overpotential curves were calculated as the difference, at the same SoC value, between the (dis)charge EMF experimentally determined by means of GITT at 25°C and the (dis)charge voltage curves measured at 25°C using the constant-current method described above.

It can be concluded from the full (dis)charge step measurements and Fig. 5.8 that the battery overpotential increases at low SoC values and remains almost constant at SoC values higher than 20%. Another important conclusion relates to the shape of the battery (dis)charge overpotentials. It follows from Fig. 5.8 that between 20 and 80% SoC the calculated (dis)charge battery overpotential is symmetrical with respect to the horizontal axis.

In summary, the overpotential measurement methods compared in sections 5.1.1 and 5.1.2 show that the build-up of the overpotential depends on the dis(charging) time, SoC, C-rate current and temperature.

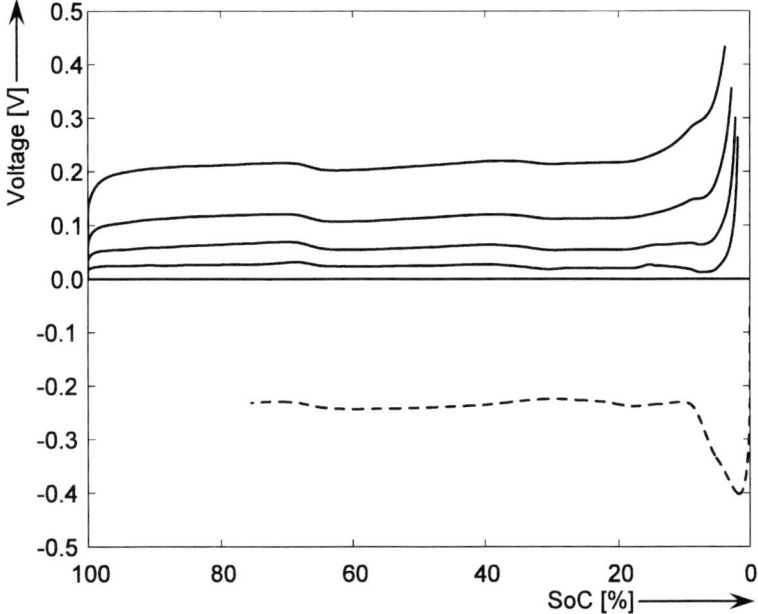

Fig. 5.8. Overpotential curves obtained at different C-rate currents (same as Fig. 5.6) at 25°C during discharge (solid lines) and charge at 1 C-rate current (dashed line) as a function of SoC.

5.2 Overpotential modelling and simulation

5.2.1 Overpotential modelling

In this section an overpotential equation inferred from previously developed physical models will be presented [1], [8]. The overpotential in which the ohmic, kinetic, diffusion overpotential and the increase in the diffusion overpotential when the battery becomes empty are considered is modelled as follows

$$\eta[V] = I \left[R_{\Omega k}(T) + R_{Ik}(I,T) + \left(R_d^0(T) + R_d^I(I,T) \right) \left(1 - e^{\frac{-c_2(T)\sqrt{t}}{\tau_d(T)}} \right) + \frac{\left(E_q^0(T) + E_q^I(I,T) \right) \left(1 - e^{-\frac{c_4(T)Q_{in}(t)}{\tau_q(T)}} \right)}{[Q_{in}(t)]^{n_0(T)+n_1(I,T)}} \right]$$

(5.1)

where $I[A] > 0$ for discharge, $I[A] < 0$ for charge, $R_{\Omega k}(T)[\Omega]$, $R_{Ik}(I,T)[\Omega] = c_0(T)[\Omega A^{-1}]I[A]$ denote the contributions of the "ohmic" and "kinetic" resistance, $R_d^0(T)[\Omega]$ and $R_d^I(I,T)[\Omega] = c_1(T)[V]/I[A]$ denote the contributions of the "diffusion" resistance, $c_2(T)$ $[s^{1/2}]$ is a constant, $\tau_d(T)[s]$ denotes the "diffusion" time constant, $n_0(T)$ (dimensionless) and $n_1(I,T) = c_5(T)[A]/I[A]$ are parameters relating to the magnitude of the diffusion overpotential, $E_q^0(T)$ [J A^{-1}] and $E_q^I(I,T)[JA^{-1}] = c_3(T)[J]/I[A]$ denote the amount of energy that cannot be extracted from the battery when the current I increases, $Q_{in}(t)$ [C] denotes the charge present in the battery at time t[s] and $c_2[s^{1/2}] = c_4[A^{-1}] = 1$ numerically. Finally, $\tau_q(T)$ is a time constant associated with the increase in overpotential in an almost empty battery in [s]. Simulation results of the mathematical overpotential implementation will be presented in the remainder of this chapter.

5.2.2 Simulation results

The results of the mathematical implementation of the overpotential function will be presented in this section. The US18500G3 Li-ion battery was used in these experiments. The overpotential model described by Eq. (5.1) needed to be fitted to the measured overpotential curves.

Fig. 5.9 illustrates the overpotentials measured and fitted at four different full discharge current C-rates as a function of the relative SoC based on the mathematical implementation of the overpotential (see Eq. (5.1)) and the measured EMF. The measurements were carried out at 25°C. In order to guide the eye, the difference between the measured and fitted overpotentials at four different discharge currents has been plotted as a function of the relative SoC in Fig. 5.10.

It can be concluded from Figs. 5.9 and 5.10 that the maximum difference between the measured and fitted overpotentials at the 0.1 C-rate discharge current was obtained at low SoC values. At 1.85% SoC the obtained difference amounted to around 57 mV. This voltage error corresponds to a low SoC error (SoC_e), i.e. $SoC_e = 0.4\%$, which can still be removed from the battery. It follows from Eq. (5.2) that even in the case of a new battery, at 25°C and at 0.1 C-rate mean discharge current this means that the SoC system will indicate 2.4 minutes more remaining run-time than is actually the case (See Eq. (5.2)).

$$t_r[\text{min}] = \frac{SoC_e[\%]}{100}\frac{1}{C_d}60 \qquad (5.2)$$

where SoC_e denotes the error in SoC in [%] and C_d denotes the mean discharge C-rate current.

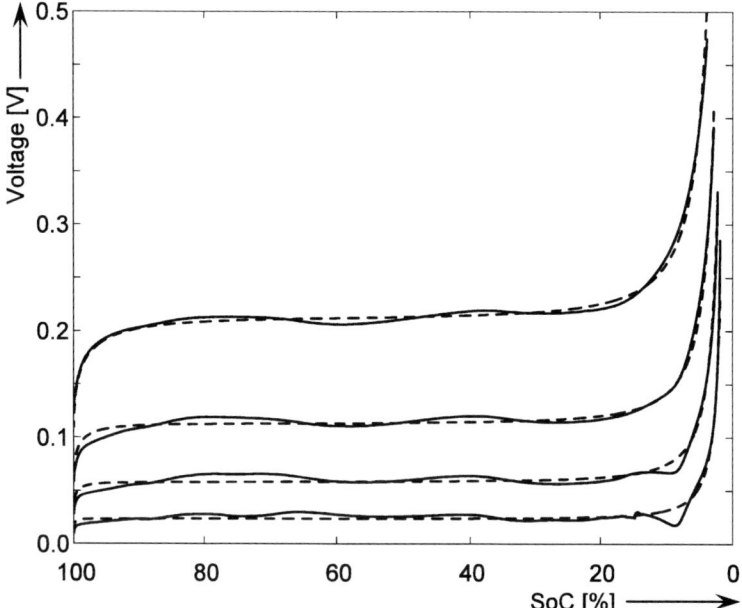

Fig. 5.9. The measured (solid lines) and the fitted (dashed lines) overpotential curves obtained at different C-rate currents and 25°C during discharge (see also Fig. 5.6). The horizontal axis shows the SoC [%] normalised to maximum capacity.

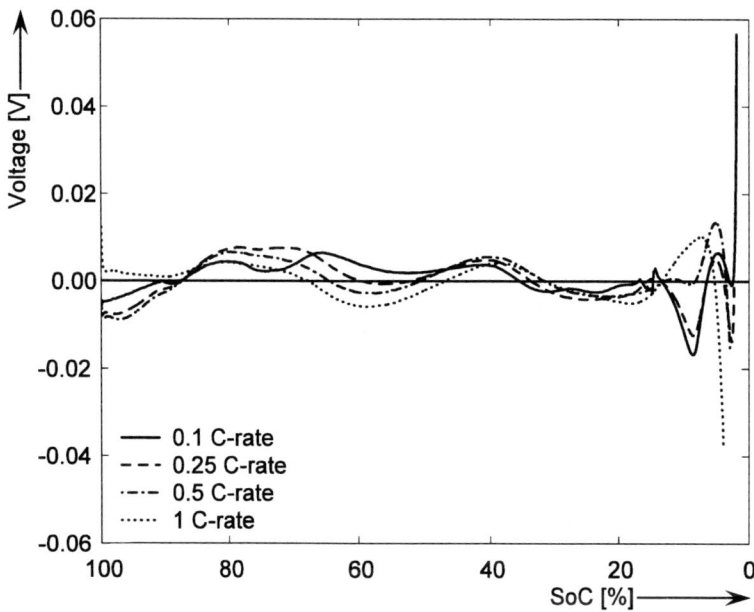

Fig. 5.10. The difference between the measured and the fitted overpotentials obtained during discharge at different C-rate currents and 25°C (see also Fig. 5.9). The horizontal axis shows the SoC [%] normalised to maximum capacity.

It follows from Figs. 5.9 and 5.10 that a very good fit (less than 10 mV difference) between the two curves is however usually obtained. Fig. 5.10 shows that the overpotential evolution can be accurately modelled for C-rate currents higher than 0.1, especially the part corresponding to low SoC values. This part of the curve is most important in practice because it describes the amount of charge remaining at a given C-rate current [1]. It can be concluded that the error generated in the overpotential implementation is small enough to allow a final accuracy within 2.4 minutes in the remaining run-time indication. More information on the overpotential implementation accuracy will be given in Chapter 7.

Based on the mathematical implementation of the overpotential described in this section, the overpotentials measured at four discharge C-rate currents have been fitted at 5, 25 and 45°C. Fig. 5.11 presents the measured and fitted battery overpotential at three different temperatures during discharge at 0.25 C-rate. In order to guide the eye, the difference between the measured and fitted overpotential at three temperatures and at 0.25 C-rate as a function of the relative SoC is plotted in Fig. 5.12.

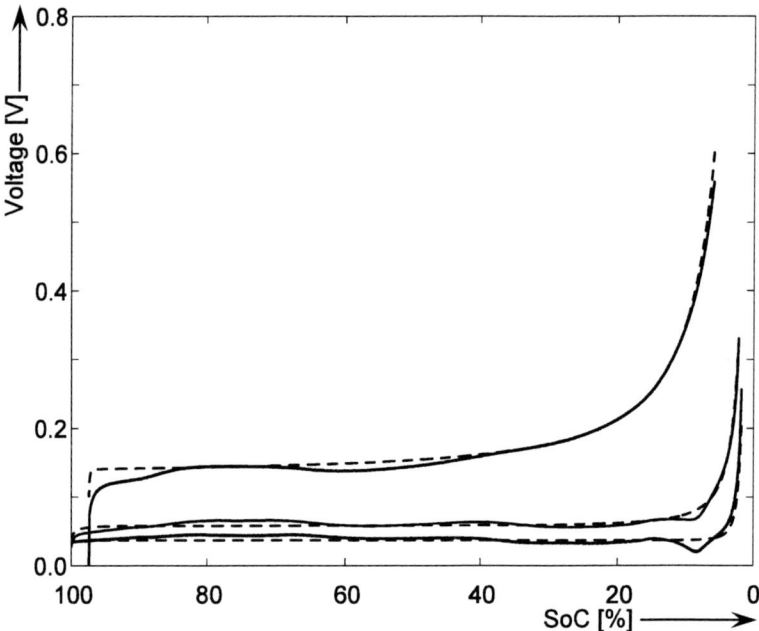

Fig. 5.11 Overpotential curves measured at 0.25 C-rate at various temperatures (solid lines; see also Fig. 5.7) and corresponding fitted curves (dashed lines) inferred from the mathematical implementation of Eq. (5.1). The horizontal axis shows the SoC [%] normalised to maximum capacity.

It follows from Figs. 5.11 and 5.12 that the maximum difference retrieved at low SoC between the measured and the fitted overpotential is obtained at 45°C. High accuracy in the overpotential modelling at low SoC is important because this part is considered for the remaining run-time calculation (see Fig. 3.4). In this situation, at 1.82% SoC the obtained overpotential difference had a value of 54 mV. This overpotential difference corresponds to a low capacity value (SoC=0.4% or 4.4 mAh), which still can be removed from the battery. Even for a new battery and at 0.25 discharging C-rate this means that the SoC system will indicate 1 minute more remaining run-time than in the real case (See Eq. (5.2)). In the majority of the cases a very good fit (under 10 mV difference) between the two curves is obtained.

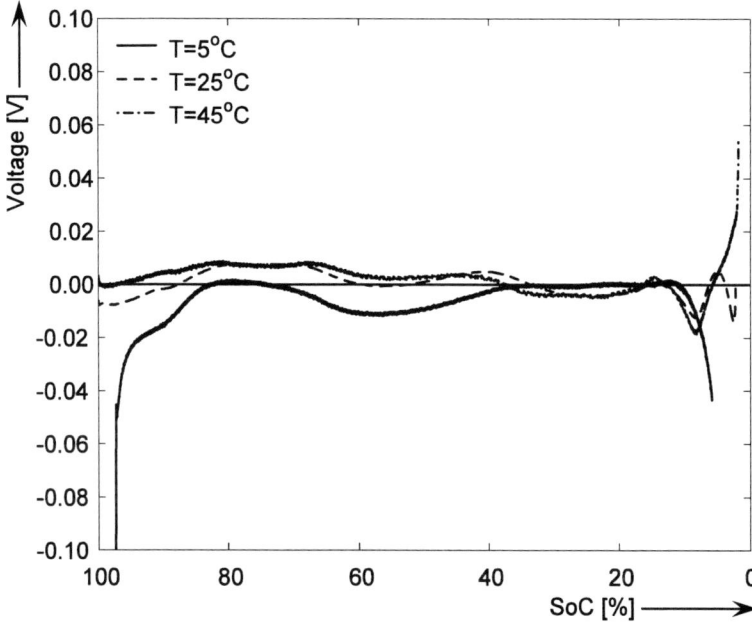

Fig. 5.12. Accuracy of the fitted overpotential curves versus the overpotential measured with the Maccor battery tester at 0.25 C-rate (see also Fig. 5.11). The horizontal axis shows the SoC [%] normalised to maximum capacity.

The overpotential model parameters values determined independently from the C-rate and used in the simulations illustrated in Figs. 5.9, 5.10, 5.11 and 5.12, respectively, are summarized in Table 5.1. Column one gives the symbol of the overpotential model parameters. The values of these parameters are given in columns two, three and four, respectively. Column five denotes the unit of the overpotential model parameters.

Table 5.1. Battery overpotential model parameter values obtained at three temperatures.

Parameter	5°C	25°C	45°C	Unit
$R_{\Omega k}$	$3.63 \ 10^{-1}$	$1.11 \ 10^{-1}$	$1.08 \ 10^{-1}$	$[\Omega]$
c_0	$-2.40 \ 10^{-1}$	$-2.02 \ 10^{-2}$	$-1.16 \ 10^{-2}$	$[\Omega A^{-1}]$
R_d^0	$1.99 \ 10^{-1}$	$9.24 \ 10^{-2}$	$4.77 \ 10^{-2}$	$[\Omega]$
c_1	$-1.13 \ 10^{-2}$	$-4.76 \ 10^{-4}$	$-7.86 \ 10^{-3}$	$[V]$
c_2	1.00	1.00	1.00	$[s^{1/2}]$
τ_d	$3.41 \ 10^{-1}$	1.05	$7.77 \ 10^{-1}$	$[s]$
E_q^0	$1.10 \ 10^{-7}$	$2.21 \ 10^{3}$	$1.00 \ 10^{6}$	$[J A^{-1}]$
c_3	$1.43 \ 10^{3}$	$5.57 \ 10^{2}$	$5.90 \ 10^{4}$	$[J]$
c_4	1.00	1.00	1.00	$[A^{-1}]$
τ_q	0.00	$5.42 \ 10^{-2}$	$3.41 \ 10^{-3}$	$[s]$
n_0	1.37	1.81	3.27	$[1]$
c_5	$2.60 \ 10^{-2}$	$1.06 \ 10^{-2}$	$2.40 \ 10^{-2}$	$[A]$

It should be noted that the scope of the method presented in this book was to achieve a high level of accuracy in SoC determination (see chapter 1). The overpotential model parameters were determined with this in mind in the fitting process, in the same way as the parameters for the EMF model (see chapter 4). For this reason these parameters are in some cases out of their physical range.

In the remainder of this section it will be shown that the parameters obtained in the discharge overpotential modelling can also be effectively used to calculate the 1 C-rate charge current overpotential. Fig. 5.13 illustrates the measured and fitted overpotentials obtained at different C-rate discharge currents and at 1 C-rate charge current. It can be concluded from this figure that the charge overpotential evolution can be accurately modelled especially in the symmetrical part between 20 and 80% SoC. More information on overpotential symmetry will be given in the next chapters of this book.

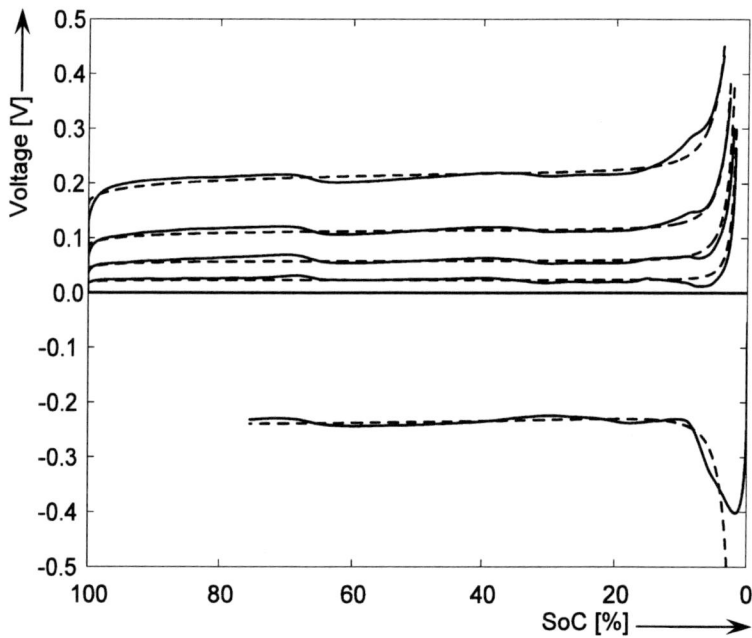

Fig. 5.13. Measured (solid lines) and fitted (dashed lines) overpotential curves obtained at different C-rate discharge currents and at 1 C-rate charge current at 25°C (see also Fig. 5.8). The horizontal axis shows the SoC [%] normalised to maximum capacity.

5.3 Conclusions

Methods for measuring and modelling battery (dis)charge overpotentials have been presented in this chapter. The measurement and modelling methods have been applied to fresh Sony US18500G3 Li-ion batteries.

Two methods for measuring battery overpotentials have been presented. In the first method the (dis)charge overpotential is obtained by means of partial (dis)charge steps at different temperatures. The EMF obtained with the aid of GITT and the newly developed voltage-relaxation model, have also been used in this method. The advantage of using this method based on partial (dis)charge steps is that the GITT measurements can be used directly to obtain information on a battery's (dis)charge overpotential.

The main drawback of this method is that insufficient information on the continuous build-up of the overpotential in time can be inferred from the duration of the (dis)charge step. This information has to be obtained using a second method involving full (dis)charge measurements. These two measurement methods show that the value of a battery's overpotential depends on the (dis)charge C-rate current, the duration of the (dis)charge step, the SoC and the temperature. As has been shown, the battery overpotential increases at low SoC and temperature values.

A new phenomenon, *i.e.* overpotential symmetry, was discovered during the analysis of the measurements results. As has been shown, the calculated (dis)charge battery overpotential is symmetrical between 20 and 80% SoC. Properties of this phenomenon will be used in the next chapters to arrive at accurate battery modelling with due allowance for the aging effect.

An overpotential model inferred from previously developed physical models has been also presented. Good fitting results have been obtained with this overpotential model. These results allow overpotential predictions with an accuracy within 0.4% SoC or 2.4 minutes' remaining run-time. The parameters obtained in the discharge overpotential fitting exercise have been used to fit the 1 C-rate charge current overpotential. It was found that the charge overpotential evolution can be accurately modelled especially in the symmetrical part between 20 and 80% SoC.

5.4 References

[1] H.J. Bergveld, W.S. Kruijt, P.H.L. Notten, Battery Management Systems, Design by Modelling, Philips Research Book Series, **1**, Kluwer Academic Publishers, Boston (2002)
[2] F.A.C.M. Schoofs, W.S. Kruijt, R.E.F. Einerhand, S.A.C. Hanneman, H.J. Bergveld, Method of and Device for Determining the Charge Condition of a Battery, US Patent 6,420,851, filed March 29 (2000)
[3] H.J. Bergvel d, H. Feil, J.R.G.C.M. Van Beek, Method of predicting the state of charge as well as the use time left of a rechargeable battery, US Patent 6,515,453, filed November 30 (2000)
[4] J.H. Aylor, A. Thieme, B.W. Johnson, A battery State-of-Charge Indicator for Electric Wheelchairs, IEEE Trans. On industrial electronics, **39**, 398–409 (1992)
[5] T. Kikuoka, H. Yamamoto, N. Sasaki, K. Wakui, K. Murakami, K. Ohnishi, G. Kawamura, H. Noguchi, F. Ukigaya, System for measuring state of charge of storage battery, US Patent 4,377,787, filed August 8 (1980)
[6] G.R. Seyfang, Battery state of charge indicator, US Patent 4,949,046, filed June 21 (1988)
[7] M.W. Verbrugge, E.D. Tate Jr., S.D. Sarbacker, B.J. Koch, Quasi-adaptive method for determining a battery's state of charge, US Patent 6,359,419, filed December 27 (2000)
[8] H.J. Bergveld, W.S. Kruijt, P.H.L. Notten, J. Power Sources, **77**, (1999) 143–158
[9] P.P.L. Regtien, F. van der Heijden, M.J. Korsten and W. Olthuis, Measurement Science for Engineers, Kogan Page Science Publisher, London (2004)

Chapter 6
Battery aging process

During a battery's lifetime, its performance or "health" tends to deteriorate gradually due to irreversible physical and chemical changes that take place with usage. To ensure accurate State-of-Charge (SoC) calculation while a battery ages, the changes that take place in a battery's overpotential and Electro-Motive Force (EMF) behaviour need to be fully understood.

This chapter presents a complete description of battery EMF and overpotential behaviour as a function of battery aging in relation to a US18500G3 Li-ion type of battery. The main aspects of battery aging will be presented in section 6.1. EMF measurement by means of GITT as a function of battery aging will be presented in section 6.2. A comparison will be made with a fresh battery's EMF. The focus in section 6.3 will be on overpotential measurements based on partial charge/discharge cycles as a function of battery aging. New adaptive systems for battery EMF and overpotential based on the phenomena discovered during analysis of the measurement results will be presented in section 6.4. These systems will enable accurate SoC calculation while a battery ages. The voltage prediction model developed in section 4.2 will also be used in the EMF adaptation method. Section 6.5 will present concluding remarks.

6.1 General aspects of battery aging

The influences of temperature and C-rate on a fresh battery's EMF and overpotential measurement and modelling have been discussed in chapters 4 and 5, respectively. To ensure accurate battery measurement and modelling, the influence of the aging effect must also be considered. General aspects of the aging of a US18500G3 Li-ion type of battery will be discussed in this section.

6.1.1 Li-ion battery aging

Various degradation processes may contribute to battery aging, *i.e.* the electrolyte decomposition, the formation of surface films on both electrodes, compromised inter-particle contact at the cathode, *etc.* [1]. In previous literature has been reported that during the battery lifetime the formation of the Co_3O_4 will also take place. This mechanism has been adopted by the authors of this book in an adaptive Li-ion model through which the aging effect has been explained. Consequently, this mechanism has been considered also in this work to explain the battery EMF aging.

The decomposition of the $LiCoO_2$ electrode in which xLi^+ ions are intercalated can be represented as follows

$$Li_xCoO_2 \rightarrow (1-x)[Co_3O_4 + O_2\uparrow]/3 + xLiCoO_2 \tag{6.1}$$

where the active electrode material decomposes into inactive Co_3O_4 material, which forms at the surface of the $LiCoO_2$ electrode and will contribute to the increase in the battery's impedance and hence overpotential, and to the decrease in maximum storage capacity.

The operational conditions determine whether an aged battery may still have acceptable remaining run-times (t_r). A battery with a high impedance may still have acceptable run-times in portable devices that require low discharge C-rate currents, *e.g.* a CD player or a wireless mouse. In the case of portable devices that require high discharge C-rate currents, *e.g.* mobile phones, digital cameras or Electrical Vehicles (EVs), the remaining run-times will on the contrary be unacceptably short. An aged battery in such a device will have to be recharged more often than when the device was new. This will lead to even more wear-out.

The formation of Co_3O_4 species at the surface of the $LiCoO_2$ electrode also contributes to the decrease in the amount of active $LiCoO_2$. In this case the number of Li^+ ions that can intercalate in the $LiCoO_2$ electrode, *i.e.* the maximum capacity of the $LiCoO_2$ electrode (Q_{max}^+), will decrease, causing the battery's maximum capacity (Q_{max}) to decrease, too. A battery with a low Q_{max} value may still have an acceptable t_r at low discharge C-rate currents, but the decrease in t_r will become appreciable at higher C-rate currents. It can be concluded that the performance of the Li-ion will deteriorate during the battery's lifetime due to the increase in impedance and/or decrease in Q_{max}.

The changing rate of a battery's impedance and maximum storage capacity is strongly dependent on the operational conditions. High C-rate (dis)charge currents, high temperatures and voltage levels during charging will speed up the degradation of these two battery characteristics. To illustrate this, the discharge capacities (Q_d) in mAh of two batteries (Q_{d1} and Q_{d2}) have been plotted as a function of cycle number in Fig. 6.1. In both examples the discharge capacity was inferred from a complete discharge step at 0.5 C-rate current by means of Coulomb counting [1]–[9].

The decrease in discharge capacity can be expressed by

$$Q_{di}[\%] = 100\left(1 - \frac{Q_{di}^j}{Q_{di}^1}\right) \quad (6.2)$$

where Q_{di} [%] denotes the decrease in Q_d [mAh] after j cycles and i the battery number.

During the operational conditions performed for battery 1 (continuous line in Fig. 6.1), the battery has been charged by means of the Constant-Current-Constant-Voltage (CCCV) method until 4.3 V at 25°C. Clearly, the battery voltage rises during the CC period until the voltage value of 4.3 V is attained. Furthermore, the current decreases in the CV-region. Charging is terminated when the current cut-off value has reached a current of 0.05 C-rate. During the CC step a 4 C-rate current has been applied. After a resting period of 30 minutes, the battery was always discharged at 0.5 C-rate. Discharging was terminated when a voltage of 3.0 V was reached. The discharge step was followed by a resting period of 30 minutes. In this case, the discharge capacity after 220 cycles was 675 mAh, whereas the initial storage capacity had been 1165 mAh. It can be concluded from this example that the discharge capacity decreased by about 42% in 220 cycles.

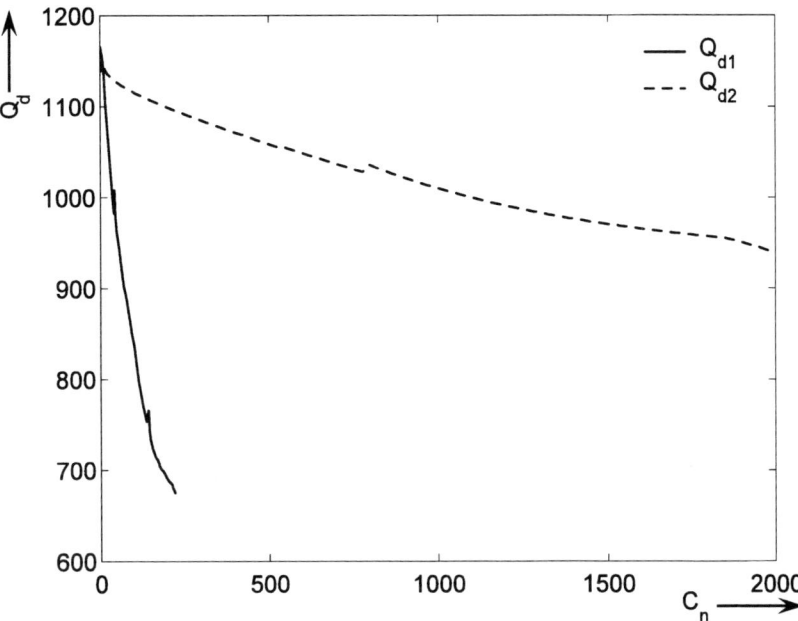

Fig. 6.1. Discharge capacity Q_d [mAh] as a function of cycle number, C_n. Q_{d1} and Q_{d2} are the discharge capacities measured for battery 1 and 2, respectively, under two different operational conditions. Battery 1 has been charged by means of the Constant-Current-Constant-Voltage (CCCV) method until 4.3 V at 25°C. Charging has been terminated when the current cut-off value has reached a current of 0.05 C-rate. During the CC step a 4 C-rate current has been applied. After a resting period of 30 minutes, the battery was always discharged until 3.0 V at 0.5 C-rate. Battery 2 has been partially (dis)charged between 30% and 70% SoC with 0.5 C-rate current at 25°C. The 40% SoC difference has been calculated as a fraction of the maximum capacity (Q_{max}) measured for a fresh battery.

In the second case (dashed line in Fig. 6.1), the battery has been partially (dis)charged between 30% and 70% SoC with 0.5 C-rate current at 25°C. The 40% SoC difference has been calculated as a fraction of the maximum capacity (Q_{max}) for a fresh battery. Each 50 cycles a complete (dis)charge has been performed. During charge the CCCV method has been applied until the 4.2 V voltage level. Similar charge termination conditions and discharging cycles with those for battery 1 have been considered. The discharge capacity after 2000 cycles was 935 mAh, whereas the initial storage capacity had been 1150 mAh. So in 2000 cycles the discharge capacity decreased 19% relative to the initial storage capacity.

It should be noted that the decrease in the discharge capacity illustrated in Fig. 6.1 is a result of two combined battery processes, *i.e.* a decrease in battery capacity and an increase in battery impedance. A correct identification and separation of these two factors will be made in next section where the battery maximum capacity will be determined independently of the battery overpotential.

6.1.2 Q_{max} measurements

For an accurate calculation of the maximum capacity of fresh and aged batteries, both 0 and 100% SoC states must be defined. In this book, 0% SoC is assumed to correspond to a battery EMF of 3 V and 100% SoC to a battery EMF of 4.175 V. The latter EMF value is based on experimental results obtained at 25°C. In repeated measurements it was observed that applying the normal CCCV charging method at 0.5 C-rate current causes a battery's voltage to relax to 4.175 V at room temperature.

In the research discussed in this book, the Q_{max} of fresh and aged batteries was determined by applying the following measurement method. The fresh batteries were first charged at a constant maximum current at a 0.5 C-rate in CC-mode until the maximum charge voltage of 4.2 V was reached in the subsequent CV-mode. The charging currents decreased in the CV-mode and the charging was ended at a predefined minimum current of 0.05 C-rate, at which the batteries were assumed to be fully charged. After a resting period of four hours, the batteries were discharged at 0.5 C-rate. Discharging was terminated when a voltage of 2.6 V was reached. The discharge step was followed by a resting period of 48 hours and a "Deep-Discharge" step at 0.001 C-rate until the battery voltage returned to 2.6 V. This "Deep-Discharge" step was followed by a resting period of 96 hours. By the end of this period the battery voltage had reached the equilibrium voltage of 3.000 V, at which the SoC level was defined to be 0%. The batteries were allowed to rest for such a long time in order to ensure that the equilibrium voltage would always be reached. The aim of the "Deep-Discharge" step at a low C-rate value was to compensate for the overpotential effect. The maximum capacity was calculated by means of cumulative Coulomb counting applied during the 0.5 and 0.001 C-rate discharging current steps.

6.2 EMF measurements as a function of battery aging

In [1]–[5] the EMF of Li-ion batteries is assumed to be dependent on aging to only a limited extent (see also chapter 3). New EMF measurement and modelling results as a function of battery aging will be presented in this section.

The accuracy of the voltage-relaxation model developed in section 4.2 as a function of battery aging will be discussed first. This model must yield highly accurate results for aged batteries to enable the influence of a not-fully-relaxed voltage on a battery's EMF to be reliably assessed. In this model, EMF measurements are obtained by applying the Galvanostatic Intermittent Titration Technique (GITT) to both fresh and aged batteries. New results obtained for EMF hysteresis as a function of battery aging will also be presented below. At the end of this section the newly developed electrochemical EMF-SoC model described in section 4.4 will be used to explain EMF dependence on battery aging.

6.2.1 The voltage-relaxation model as a function of battery aging

A new voltage-relaxation model was developed in section 4.2 of this book. It was shown that this model can be used to accurately predict a battery's equilibrium voltage on the basis of only the first few minutes of the voltage-relaxation curve. Accurate results have been obtained by applying the voltage-relaxation model to fresh batteries under a large range of conditions. As mentioned in section 4.2, the model parameters are calculated and adapted on-line. This means that the voltage-relaxation model should provide accurate results for aged batteries, too. This is important with respect to accurate interpretation of EMF measurements obtained for aged batteries. Results obtained with the voltage-relaxation model as a function of battery aging at different SoCs and temperatures will also be presented in this section. Voltage-relaxation results obtained for a fresh battery will be compared with voltage-relaxation results obtained for a 42% Q_{dd} battery (see section 6.1). The latter battery was chosen for the assessment of the accuracy of the voltage-relaxation model on account of its more pronounced aging effect.

Fig. 6.2 analyses the voltage-relaxation process as a function of battery aging by comparing the Open-Circuit Voltage (OCV) of a fresh battery OCV_f with that of an aged battery OCV_a measured after a discharge step at the same SoC, C-rate current and temperature, *i.e.* 28% SoC, 0.1 C-rate current and 25°C, respectively.

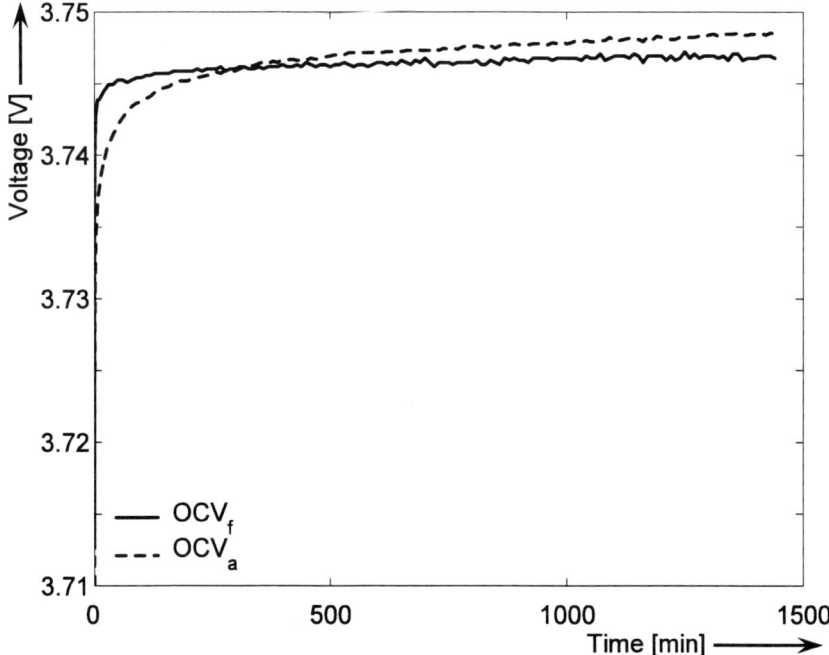

Fig. 6.2. The relaxation processes of a fresh and an aged battery after a discharge step at 28% SoC, 0.1 C-rate current and 25°C as a function of time in [min.].

It follows from Fig. 6.2 that after a discharge step a battery's OCV doesn't coincide with its equilibrium voltage, *i.e.* EMF, during the relaxation process. The value of OCV_f changes from 3.712 V immediately after the current interruption to about 3.746 V after 1440 minutes, whereas the value of OCV_a changes from 3.663 V immediately after the current interruption to about 3.748 V after 1440 minutes. It can be observed in Fig. 6.2 that in our experiment OCV_f became constant after about 500 minutes, whereas OCV_a reached the EMF after about 1200 minutes. A battery is assumed to be in equilibrium when the difference between the battery voltage measured during the relaxation process and the EMF voltage after 1440 minutes is lower than 1 mV.

It can be concluded from Fig. 6.2 that the OCV evolution during the relaxation process of a fresh battery differs from that during the relaxation process of an aged battery. So when the same parameters are assumed for a fresh and an aged battery in the voltage-relaxation model the voltage prediction result will be inaccurate. It should be noted that the OCV evolution during the relaxation process also provides information on a battery's overpotential. More information on the overpotential will be given further on in this chapter.

The course of the SoC error (SoC_e) calculated by means of voltage prediction (V_p), $SoC_e(V_p)$ for different SoCs and temperatures will be further considered in assessing the accuracy of the voltage-relaxation model. For the purpose of comparison, we will also consider the SoC error calculated by means of OCV, $SoC_e(OCV)$. The accuracy of the voltage-relaxation model was calculated using the SoC_e determination method described in section 4.2. In this method, for each moment during the relaxation process, V_p and OCV are converted into an SoC value by using a reference EMF-SoC curve. $SoC_e(V_p)$ was therefore compared with $SoC_e(OCV)$ throughout the entire relaxation process. Fig. 6.3 shows the courses of $SoC_e(V_p)$ and $SoC_e(OCV)$ considered after a discharge step at 0.1 C-rate current and 5°C as a function of time.

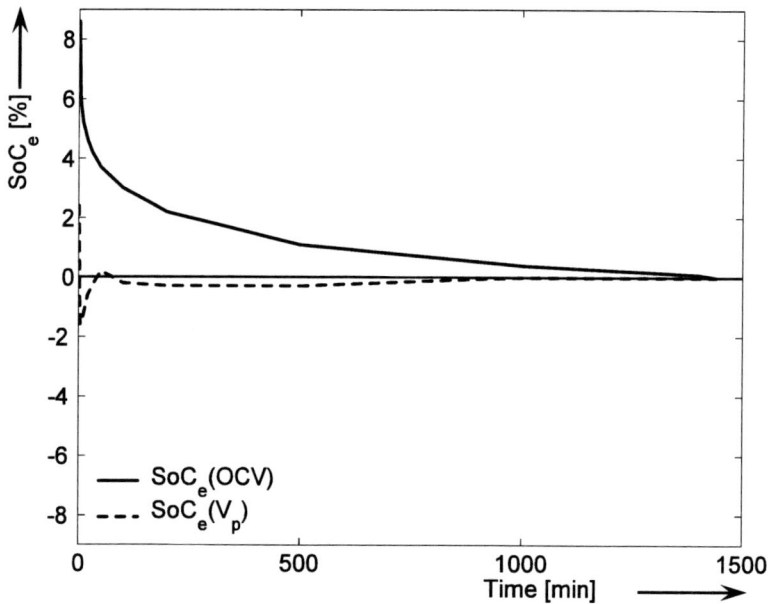

Fig. 6.3. Error in SoC calculated on the basis of the predicted voltage, $SoC_e(V_p)$, compared with the SoC error calculated on the basis of the battery's OCV, $SoC_e(OCV)$, considered after a discharge step at 32.4% SoC, 0.1 C-rate current and 5°C. The horizontal axis shows the relaxation time in [min.].

It follows from Fig. 6.3 that the SoC calculated by means of voltage prediction, $SoC(V_p)$ after two minutes of rest is close to the final SoC value calculated on the basis of the EMF voltage after full relaxation. The calculated SoC error, $SoC_e(V_p)$, is –0.4%, whereas the SoC error calculated on the basis of the battery's OCV, $SoC_e(OCV)$, is 7.3%. From this example it can be concluded that the SoC accuracy calculated on the basis of the predicted voltage after two minutes of rest is equal to the SoC accuracy calculated on the basis of the OCV considered after 1000 minutes of rest. The system based on the SoC-EMF relationship and voltage-relaxation model works 500 times faster in this situation. It can also be concluded from Fig. 6.3 that for the first 1400 minutes the SoC accuracy calculated on the basis of the predicted voltage is better than that inferred from the OCV. After this rest period both SoC indications are about equally accurate.

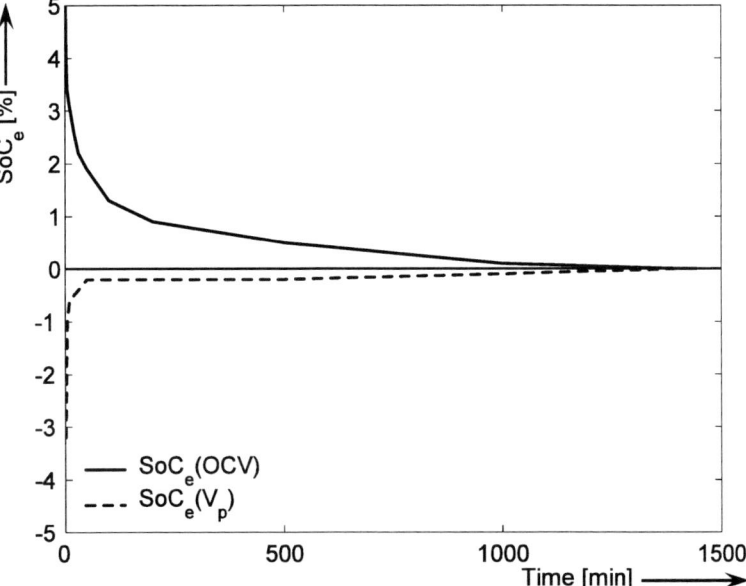

Fig. 6.4. $SoC_e(V_p)$ compared with $SoC_e(OCV)$ after a discharge step at 32% SoC, 0.1 C-rate current and 25°C as a function of time in [min.].

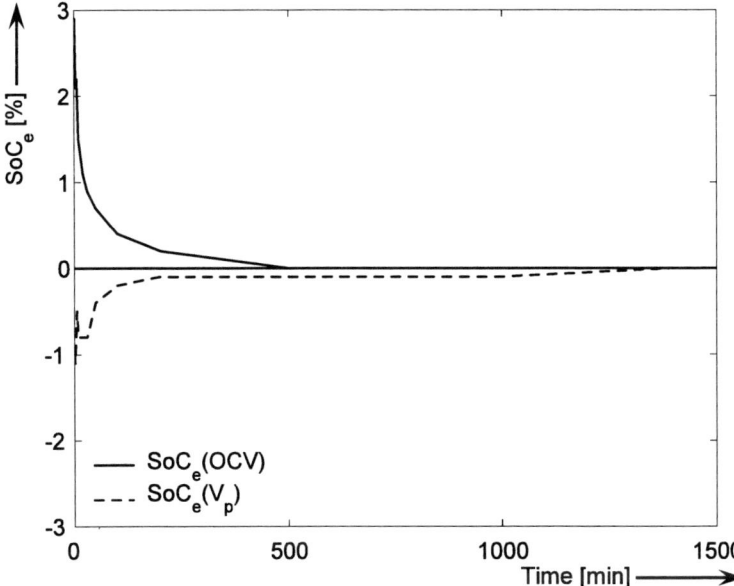

Fig. 6.5. $SoC_e(V_p)$ compared with $SoC_e(OCV)$ after a discharge step at 32% SoC, 0.1 C-rate current and 45°C as a function of time in [min.].

The courses of $SoC_e(V_p)$ and $SoC_e(OCV)$ measured after a discharge step of 0.1 C-rate current, at 25 and 45°C are plotted in Figs. 6.4 and 6.5, respectively. It follows from Fig. 6.4 that $SoC(V_p)$ after five minutes of rest is close to the final SoC value calculated on the basis of the EMF voltage after full relaxation. In this case V_p leads to a –1% SoC_e whereas OCV leads to a 3.4% SoC_e. The latter error is reduced to 1% after a rest period of 200 minutes. So the system based on the SoC-EMF relationship and voltage-relaxation model works 40 times as fast.

Similar conclusions hold for the situation measured at 45°C, as shown in Fig. 6.5. In this case $SoC_e(V_p)$ is –1% after one minute of rest. $SoC_e(OCV)$ is 2.9% and 1% after a rest period of 20 minutes: a 20-fold improvement in "speed".

The courses of $SoC_e(V_p)$ and $SoC_e(OCV)$ measured after a charge step at 0.1 C-rate current and 5°C as a function of time are plotted in Fig. 6.6.

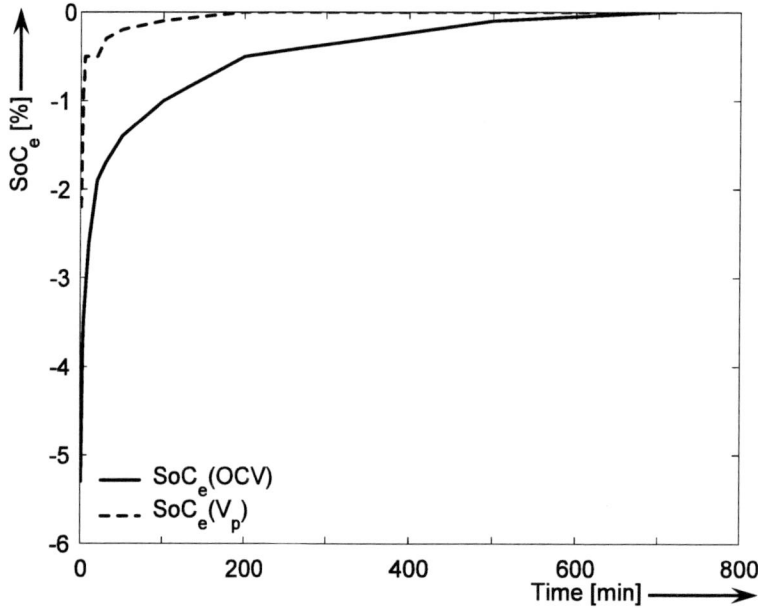

Fig. 6.6. $SoC_e(V_p)$ compared with $SoC_e(OCV)$ after a charge step at 67% SoC, 0.1 C-rate and 5°C as a function of time in [min.].

Fig. 6.6 shows that $SoC(V_p)$ calculated after a five minutes' rest leads to an SoC value close to that based on the EMF voltage. $SoC_e(V_p)$ is –0.5% in this case, whereas $SoC_e(OCV)$ is –3.3%. An SoC error of –0.5% is obtained when the battery's OCV is considered after a rest period of 200 minutes. So the system works 40 times faster.

A phenomenon similar to that illustrated in Fig. 6.6 can be seen in Figs. 6.7 and 6.8, representing the situation at 25 and 45°C, respectively. In Fig. 6.7, for example, the value of $SoC(V_p)$ after one minute's rest is close to the final SoC calculated on the basis of the EMF voltage. The $SoC_e(V_p)$ calculated after this rest period is –0.5%, whereas the SoC error calculated on the basis of the OCV $SoC_e(OCV)$ is –4.9%. An SoC error of 0.5% is obtained when the OCV is considered after a rest period of 200 minutes. The "speed" is increased by a factor of 200. Similarly, in the example illustrated in Fig. 6.8, an SoC accuracy to within 0.4% is obtained after two minutes of relaxation on the basis of V_p and the system's "speed" based on the SoC-EMF relationship and voltage-relaxation model is improved 250 times.

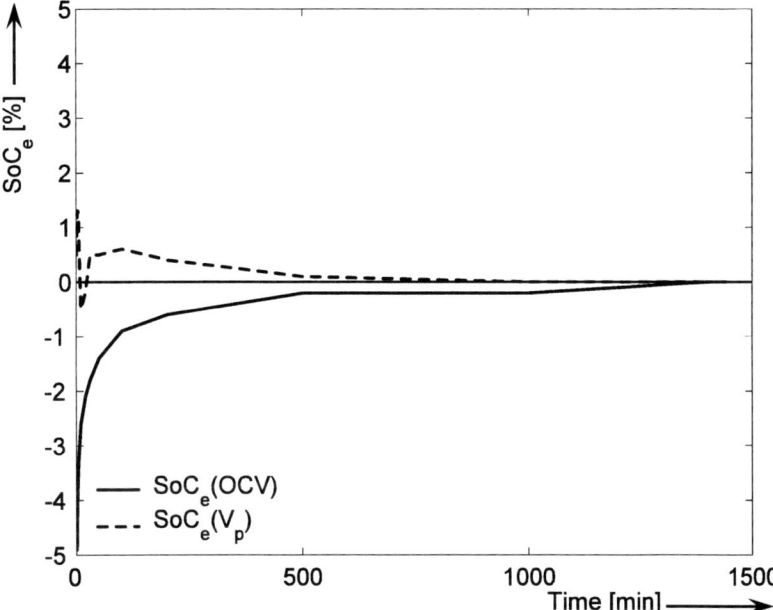

Fig. 6.7. $SoC_e(V_p)$ compared with $SoC_e(OCV)$ after a charge step at 48% SoC, 0.1 C-rate and 25°C as a function of time in [min.].

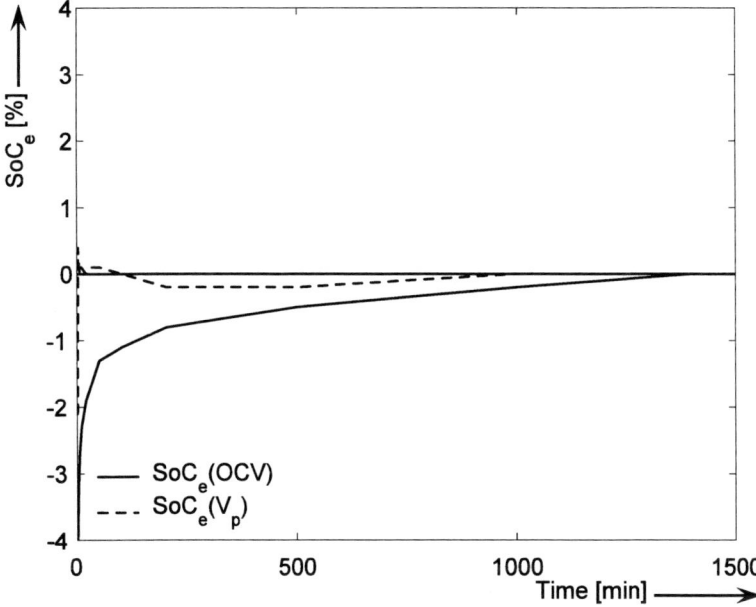

Fig. 6.8. $SoC_e(V_p)$ compared with $SoC_e(OCV)$ after a charge step at 48% SoC, 0.1 C-rate and 45°C as a function of time in [min.].

In summary, the SoC of an aged battery calculated on the basis of the predicted voltage after a rest period of five minutes at different SoCs and temperatures is always accurate to within 1%. The voltage-relaxation model will therefore be used for a correct interpretation of the charge/discharge EMF obtained by means of GITT.

6.2.2 EMF GITT measurement results obtained for aged batteries

In order to accurately determine the influence of battery aging on the EMF, Q_{max} should first be calculated. This calculation is necessary for a correct definition of SoC. Therefore, the Q_{max} experiment presented in section 6.1 was carried out using fresh and aged Li-ion batteries. It was observed that although the aged batteries were subjected to a "Deep-Discharge" step at a low C-rate value, the increase in the batteries' overpotential still influenced the EMF values, *i.e.* 3.000 and 4.175 V, at 0 and 100% SoC, respectively. As this influence was observed mainly at low and high temperatures, *i.e.* 5 and 45°C, respectively, the measurements were adjusted accordingly. For instance, if the battery EMF was found to reach a voltage level higher than 3 V, additional "Deep-Discharge" steps were applied after the 96 hour relaxation period. A similar correction was applied in the case of charging, by applying additional (dis)charging steps in order to obtain the EMF level of 4.175 V.

Let's take as an example of correction through calculation a battery EMF of 4.170 V after CCCV charging and a 4 hours' resting period. In this example a 5 mV voltage and corresponding capacity have to be added to the measured EMF and capacity values. The capacity corresponding to the 5 mV EMF value can be inferred from previous measurements of the capacity-EMF relationship between 4.170 and 4.175 V. A similar correction through calculation can be applied for discharging.

The described maximum capacity measurement yielded a Q_{max} value of 1173 mAh for a fresh battery. This value will be regarded as the maximum reference capacity, Q_{maxr} below. The aged batteries discussed in section 6.1 have a Q_{max1} = 875 mAh and Q_{max2} = 1110 mAh. The differences between the measured Q_{max1} and Q_{d1} values and between the measured Q_{max2} and Q_{d2} values, 200 mAh and 175 mAh, respectively, are explained by the influence of the battery overpotential in the Q_d measurement described in section 6.1. The capacity loss is calculated using Eq. (6.3). The capacity losses in aged batteries amount to Q_{loss1} = 25.4% and Q_{loss2} = 5.4%. It should be noted that the Q_{loss1} value is lower than the acceptable battery capacity defined in the cycle life, *i.e.* 20% (see section 2.5). This battery will therefore be assumed to represent a worst-case scenario in the measurements and tests presented in this book.

$$Q_{loss}[\%] = 100 - 100 \frac{Q_{max}}{Q_{maxr}} \qquad (6.3)$$

After the maximum battery capacity had been accurately determined, GITT measurements were carried out in order to determine the EMF as a function of battery aging. The following measurement method was applied. First, a battery was charged from 3 V at 0.1 C-rate in 25 steps. For each step a limit of 4.3 V in voltage and 4% SoC increase was assumed. Each charge step was followed by a rest period. The rest period was chosen as a function of SoC. At low SoCs long relaxation times of 24 hours are required, and at higher SoCs shorter relaxation times of 12 hours are already sufficient. Next, 25 discharge steps of 4% SoC were induced (0.1 C-rate) until the voltage limit of 2.6 V was reached. Each discharge step was again followed by a rest period. The rest period was also chosen as a function of SoC. In this way, 25 measured EMF data points were obtained for both charging and discharging. The low C-rate current was chosen in order to obtain an equilibrium voltage faster. The experiment was carried out at 5, 25 and 45°C.

Especially at low SoCs and temperatures, the battery voltage had not reached the EMF equilibrium voltage by the end of the rest periods (see also chapter 4). For this reason, the voltage-relaxation model presented in sections 4.2 and 6.2 was used to predict the EMF voltage. The voltages measured during the rest periods were used as input in the voltage-relaxation model. In this example, a 96 hours' rest period was used for the voltage prediction for practical reasons (see sections 4.2 and 6.2).

The 25 EMF points obtained by means of the GITT have been fitted according to a mathematical equation. It should be specified that the fitted EMF curve passes through the 25 measured points. Fig. 6.9 presents the EMF inferred by means of GITT as a function of aging during the discharge cycles at 25°C. The discharge EMF obtained for a fresh battery (EMF_f) was compared with the discharge EMFs obtained for the aged batteries with 5.4% ($EMF_{a5.4\%}$) and 25.4% ($EMF_{a25.4\%}$) capacity loss. The differences between the EMFs obtained for a fresh battery and the aged batteries are plotted in Fig. 6.10.

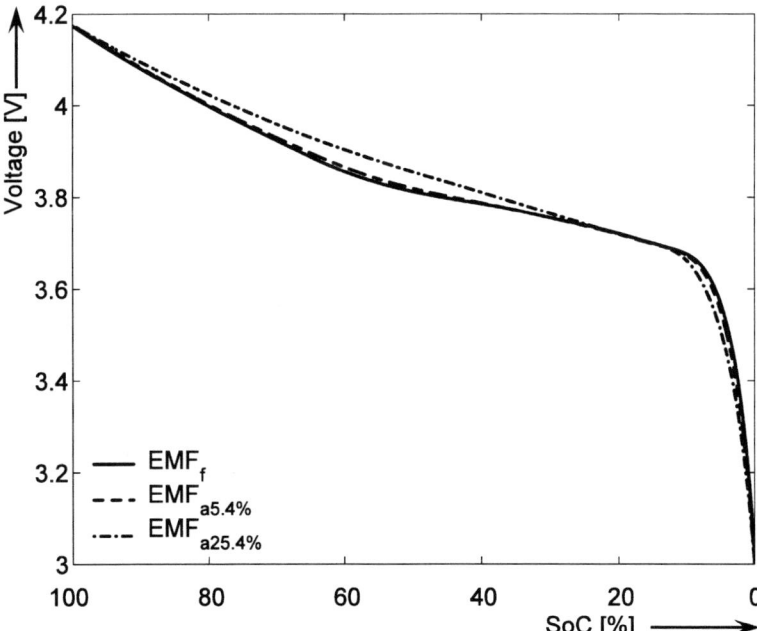

Fig. 6.9. EMFs for fresh (f) and aged (a) batteries during discharging at 25°C as function of SoC. $EMF_{a25.4\%}$ corresponds to battery 1 and $EMF_{a5.4\%}$ corresponds to battery 2. The EMF measurements have been obtained by means of Galvanostatic Intermittent Titration Technique (GITT) for both fresh and aged batteries. The maximum capacity for both fresh and aged batteries equals the capacity taken out from the battery during a complete discharge cycle performed between the 4.175 and 3.0 V EMF levels.

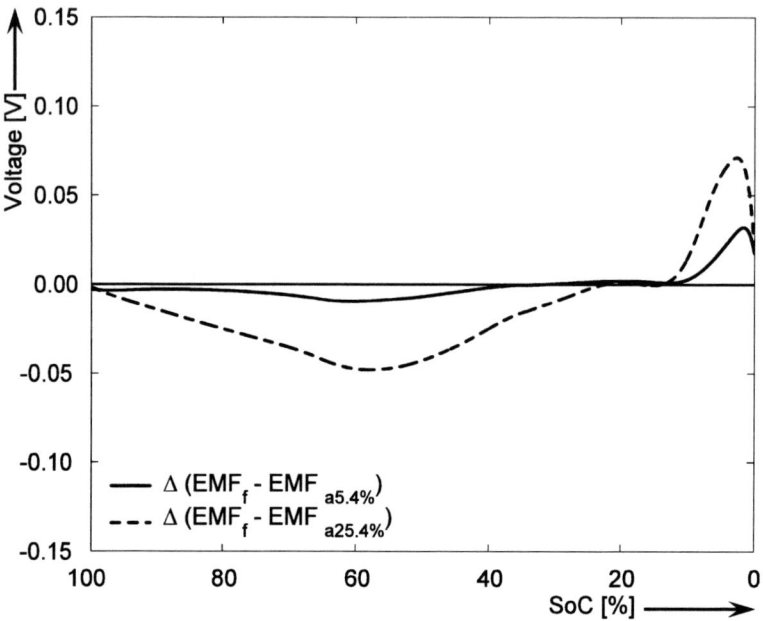

Fig. 6.10. Differences in EMF between fresh (*f*) and aged (*a*) batteries during discharging at 25°C as a function of SoC [%].

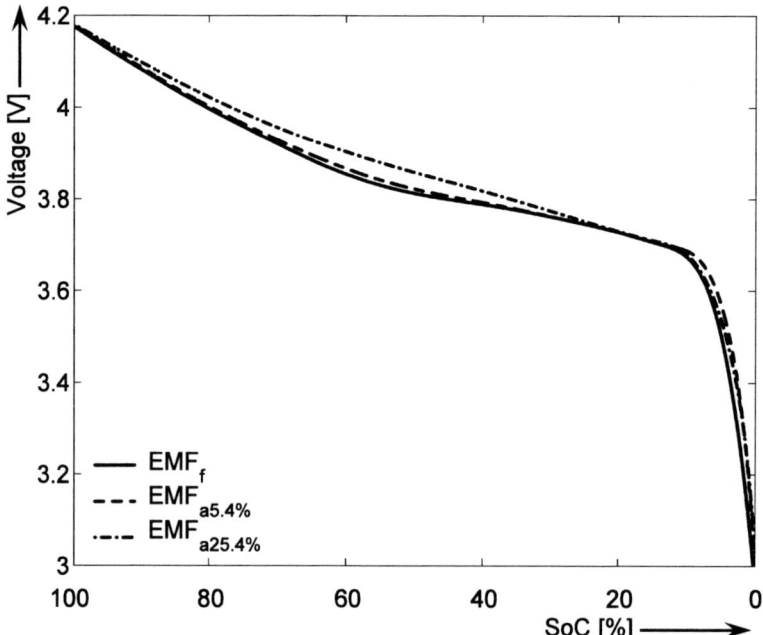

Fig. 6.11. EMFs obtained for fresh (*f*) and aged (*a*) batteries during discharging at 5°C as a function of SoC [%].

According to Figs. 6.9 and 6.10, the difference between the fresh and aged battery EMF increases with aging. For instance, a difference of −48 mV was obtained at 57.4% SoC. This means that when the EMF is used without taking into consideration the aging effect by modelling only EMF_f, the SoC indication system based on the EMF will yield an SoC value of 57.4% for an aged battery (a25.4%), whereas the SoC value is actually 46.9%, implying an inaccuracy of –10.5%.

A possible explanation for the EMF dependence on battery aging will be given in section 6.2.4. Figs. 6.11 and 6.13 show the EMF dependencies obtained for fresh and aged batteries during the discharge cycle at 5 and 45°C, respectively. The differences between the "fresh" and "aged" EMFs at 5 and 45°C are plotted in Figs. 6.12 and 6.14, respectively.

These figures show that at all temperatures a similar increase in EMF differences was observed when the batteries aged. For instance, a difference of about –48 mV was obtained at 57.4% SoC at all the considered temperatures. The small difference between the EMFs measured at different temperatures and low SoCs is explained by the small number of interpolation points chosen at low SoCs in the experiments. It can be concluded that the chosen GITT measurement method does not influence the EMF determination during discharge.

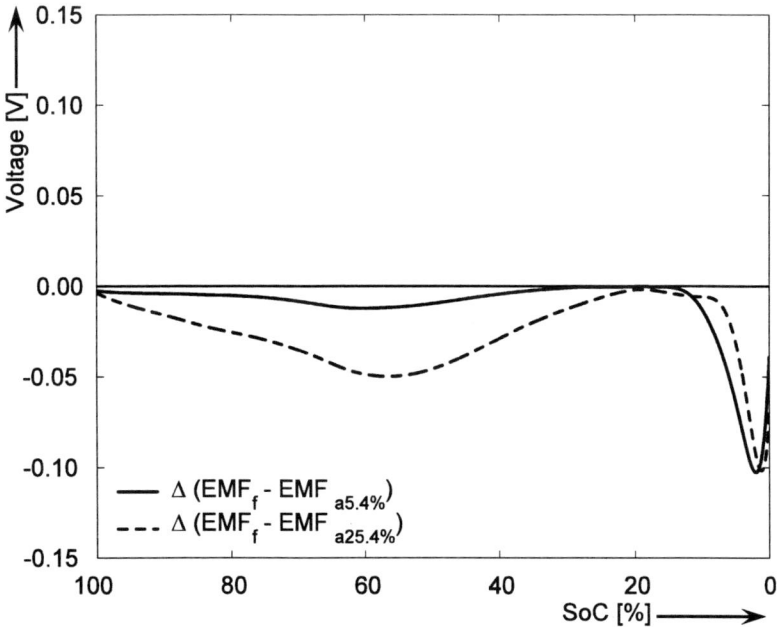

Fig. 6.12. Differences in EMF between fresh (*f*) and aged (*a*) batteries during discharging at 5°C as a function of SoC [%].

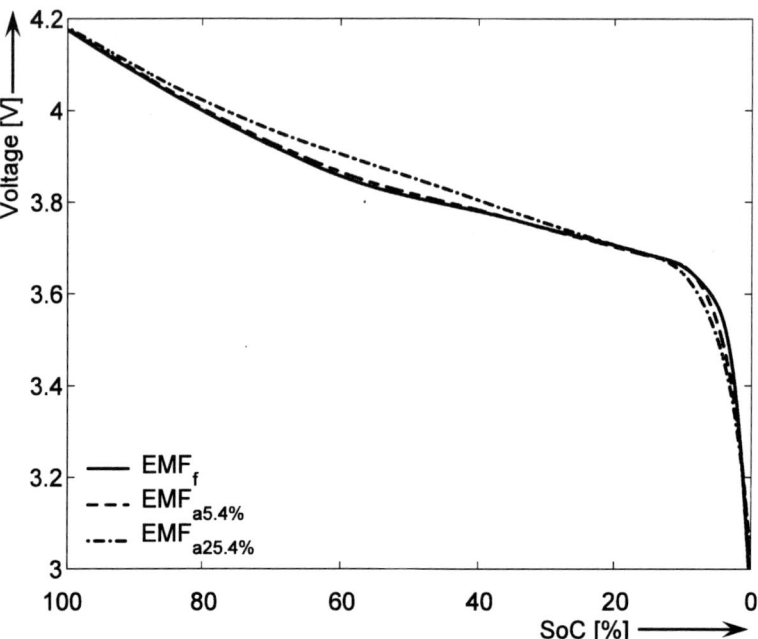

Fig. 6.13. EMFs obtained for fresh (*f*) and aged (*a*) batteries during discharging at 45°C as a function of SoC [%].

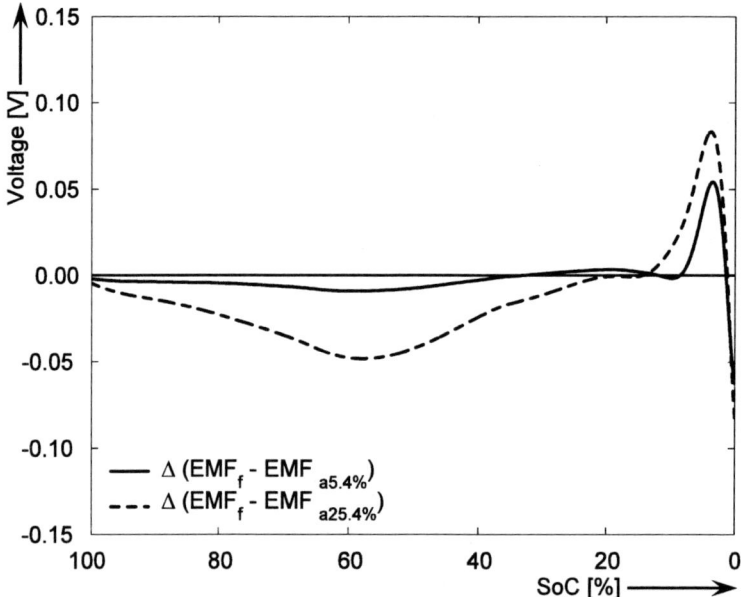

Fig. 6.14. Differences in EMF between fresh (*f*) and aged (*a*) batteries during discharging at 45°C as a function of SoC [%].

6.2.3 The charge/discharge Electro-Motive Force difference as a function of battery aging

So far, the EMF measured by means of GITT has been considered during discharge only. However, as shown in chapter 4, the discharge and charge EMF curves differ due to the so-called hysteresis effect. Therefore, EMFs measured during charging will be discussed in this section.

Fig. 6.15 compares the charge EMF measured at 25°C in a fresh battery EMF_f with the EMF of the battery with 5.4% capacity loss, $EMF_{a5.4\%}$, and that obtained for the 25.4% capacity loss battery $EMF_{a25.4\%}$. The difference between the EMFs obtained for the fresh battery and the aged batteries is plotted in Fig. 6.16.

It can be concluded from Figs. 6.15 and 6.16 that the maximum difference between EMF_f and $EMF_{a5.4\%}$ is 60 mV at 3.1% SoC. This leads to an SoC error of –0.8% when aging of the EMF curve is not considered. The measurement described above was repeated at 5 and 45°C. The results are illustrated in Figs. 6.17–6.20.

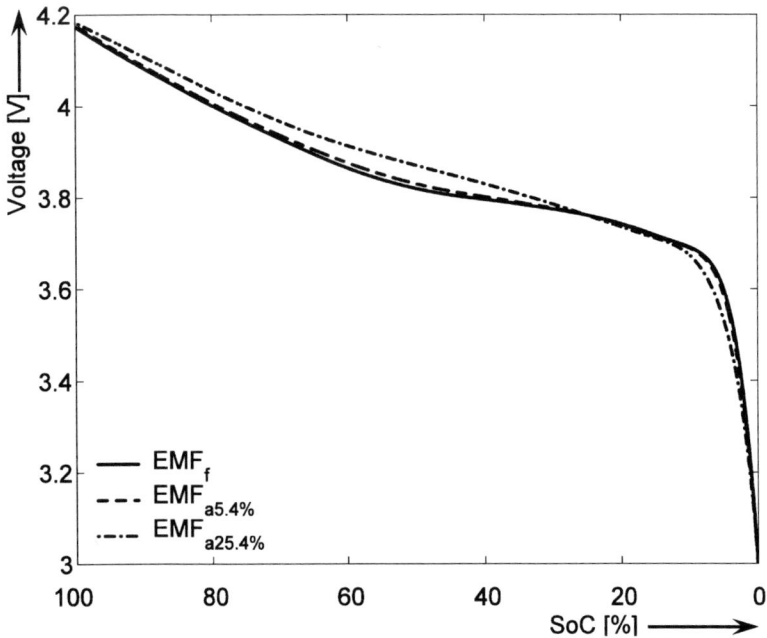

Fig. 6.15. The charge EMFs at 25°C obtained for fresh (*f*) and aged (*a*) batteries. The horizontal axis shows the SoC [%] normalised to maximum capacity.

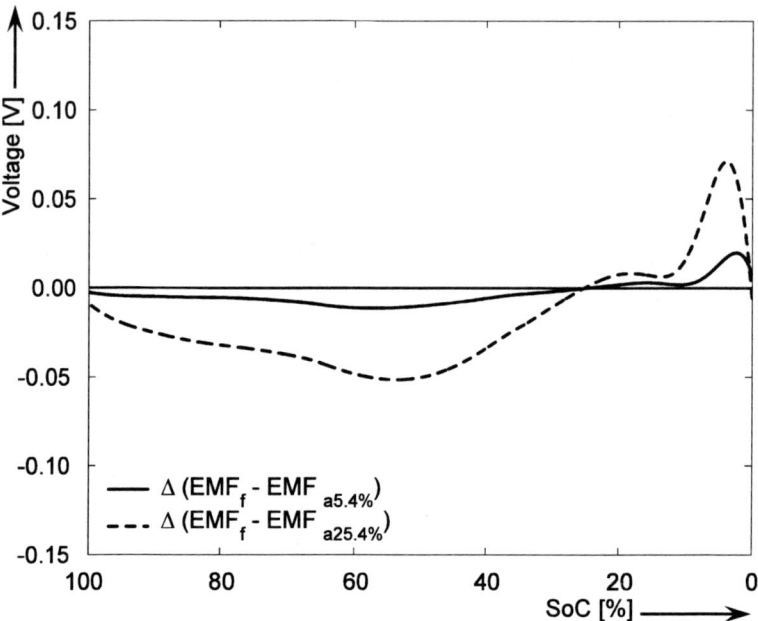

Fig. 6.16. The differences in EMF between fresh (*f*) and aged (*a*) batteries during charging at 25°C plotted as a function of SoC [%] normalised to maximum capacity.

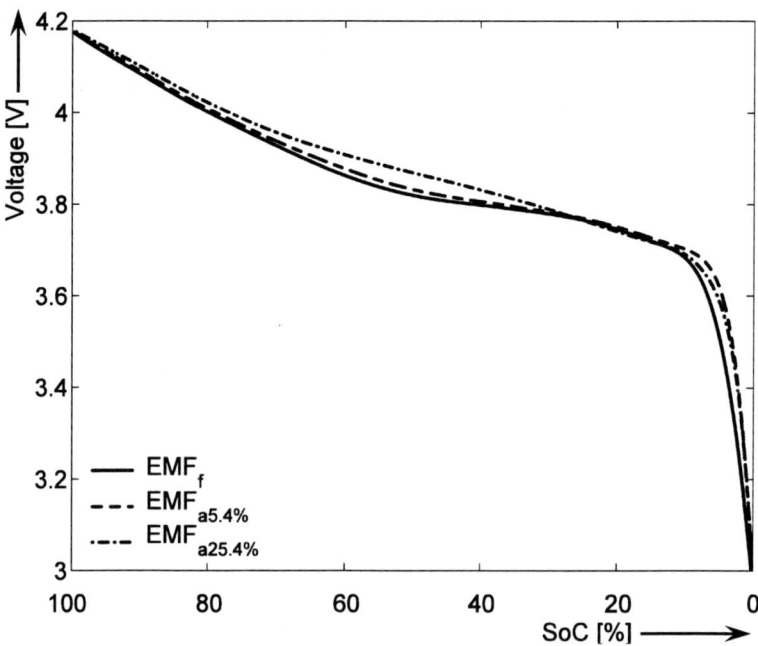

Fig. 6.17. The charge EMF obtained at 5°C EMF for fresh (*f*) and aged (*a*) batteries. The horizontal axis shows the SoC [%].

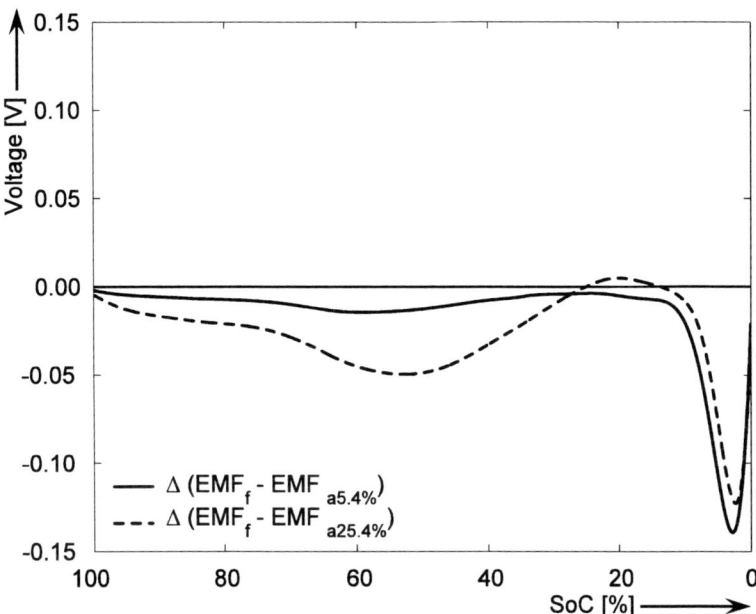

Fig. 6.18. Differences in EMF between fresh (*f*) and aged (*a*) batteries during charging at 5°C plotted as a function of SoC [%] normalised to maximum capacity.

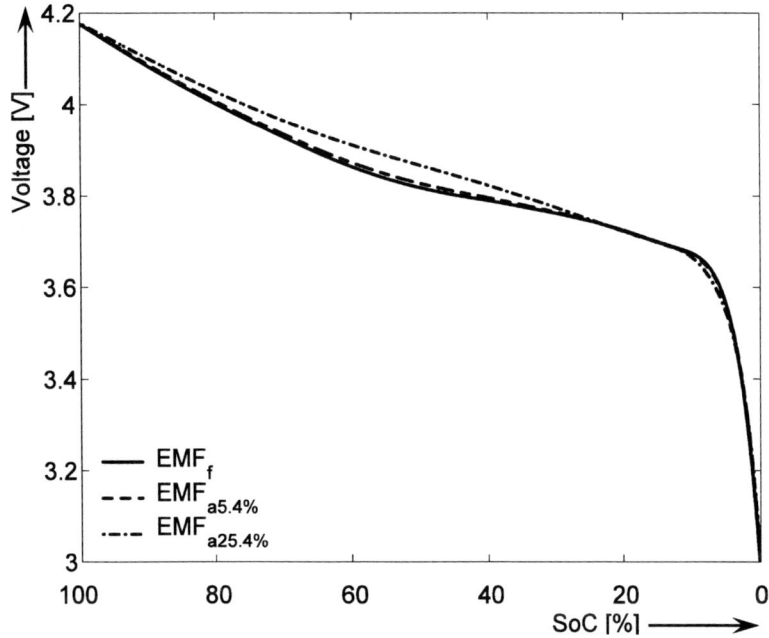

Fig. 6.19. The charge EMF obtained at 45°C for fresh (*f*) and aged (*a*) batteries. The horizontal axis shows the SoC [%].

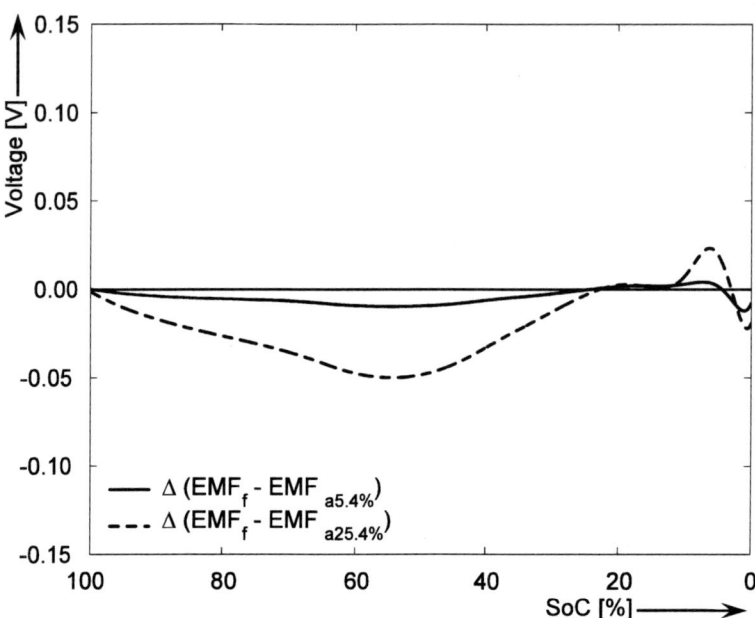

Fig. 6.20. Differences in EMF between fresh (*f*) and aged (*a*) batteries during charging at 45°C plotted as a function of SoC [%] normalised to maximum capacity.

It can be concluded from Figs. 6.15–6.20 that a maximum difference of −139 mV was obtained between the fresh and the 5.4% capacity loss battery at 5°C and around 2.8% SoC. This leads to an SoC error of −1.7% when aging of the charge EMF curve is not considered. Moreover, the difference between the fresh and aged battery EMFs was consistently the same at all temperatures. For example, a difference of about −49 mV was obtained between EMF_f and $EMF_{a25.4\%}$ at 57.4% SoC for all three temperatures.

The EMF at 25°C obtained for fresh and aged batteries during charging and discharging is illustrated in Fig. 6.21. The difference between the corresponding EMF curves is plotted in Fig. 6.22. A maximum EMF difference of 39 mV was obtained for the 5.4% capacity loss battery at 4.7% SoC. Ignoring this EMF difference leads to an error of −0.7% SoC. This effect will be more pronounced in the flat region of the EMF-SoC curve, where even small differences in EMF will cause substantial errors in SoC. For instance, a difference of about 24 mV was obtained for the EMF measured during the charge and discharge cycles in the case of the 5.4% capacity loss battery at 28.0% SoC. In this case ignoring the difference between the charge and discharge EMF will lead to an error in SoC of −6.7%.

The difference between the fresh and aged EMFs was moreover found to be consistently the same at all temperatures. It should also be noted that in GITT and voltage-relaxation measurement methods battery voltage is also predicted to ensure accurate EMF determination. The chosen measurement method did therefore not influence the determination of the EMF of the fresh and aged batteries. The EMF difference between charge and discharge (see Figs. 6.21 and 6.22) may be explained by hysteresis (*H*), which may be introduced by the $LiCoO_2$ electrode. A possible cause of the hystereses could be phase transitions (*ph*). For further information on electrochemical hysteresis the reader is referred to [10]–[12].

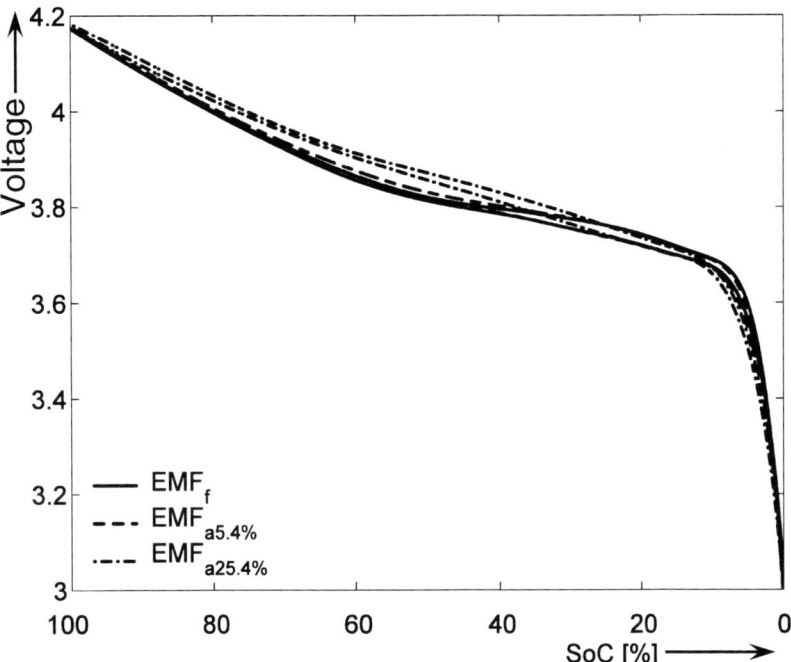

Fig. 6.21. The charge/discharge EMF obtained at 25°C for fresh (*f*) and aged (*a*) batteries. The horizontal axis shows SoC [%] normalised to maximum capacity.

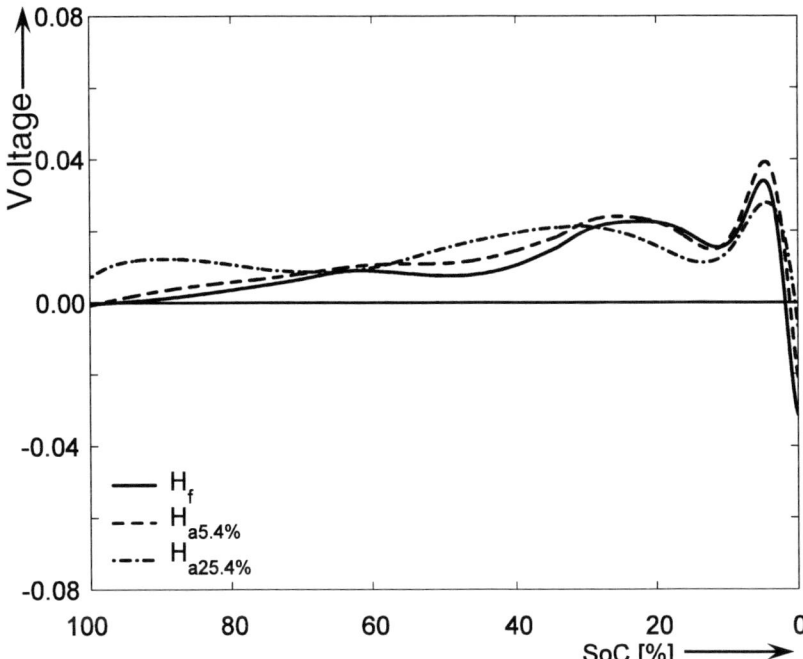

Fig. 6.22. The charge/discharge EMF difference (H) obtained at 25°C as a function of battery aging. The horizontal axis shows SoC [%] normalised to maximum capacity.

From what has been said above it can be concluded that the charge/discharge EMF dependence on the aging effect should be considered to ensure accurate SoC determination.

6.2.4 EMF modelling as a function of battery aging

A physical model for the EMF-SoC relationship in fresh batteries was presented in section 4.4. This model can be used to calculate the SoC corresponding to a given EMF and temperature. As discussed above, the measured EMF curve is however also dependent on battery aging. In this section the physical EMF model developed in section 4.4 will be used to explain the EMF dependence on battery aging in qualitative terms.

As expressed in Eq. (6.1), the $LiCoO_2$ electrode will decompose during aging and an inactive layer may be formed at the $LiCoO_2$ electrode surface. The amount of cyclable electrochemically active Li^+ ions in the battery will consequently decrease. For this reason, the EMF model parameters relating to the battery's design (Fig. 4.19) will also change, whereas the EMF parameters relating to the battery's thermodynamics will remain unchanged. This situation is schematically represented in Fig. 6.23, in which the aged Li-ion battery is characterised by the same set of parameters as the fresh battery (Fig. 4.19). The defined parameters relating to the design of a fresh battery are also represented to enable comparison.

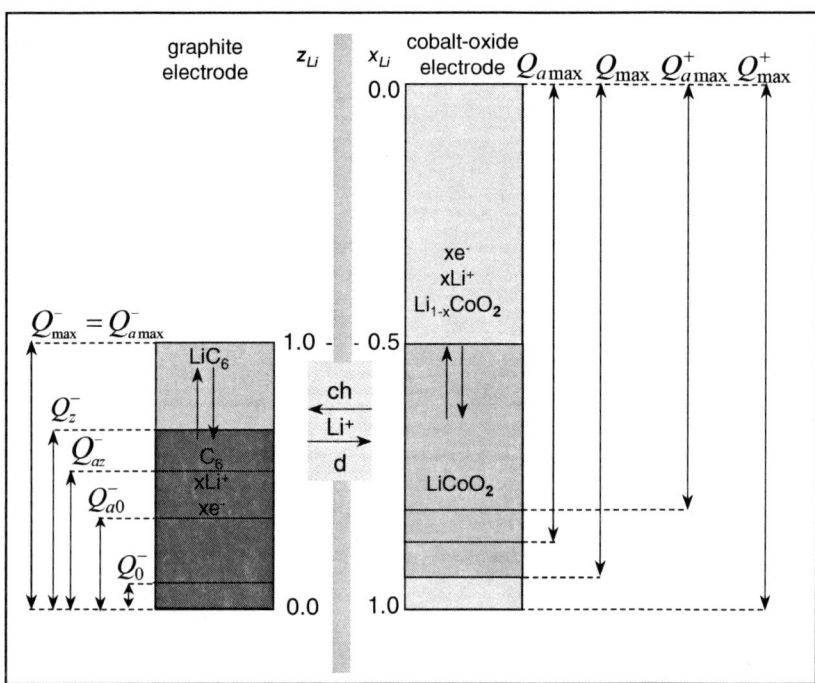

Fig. 6.23. Schematic representation of the EMF-SoC relationship parameters relating to the design of fresh and aged batteries.

Battery aging process

It follows from Fig. 6.23 that the amount of electrochemically active Li^+ ions inside an aged battery $Q_{a\,max}$ exceeds the maximum capacity of the positive electrode $Q^+_{a\,max}$. This situation can be explained by the decomposition of the $LiCoO_2$ electrode or the formation of a Co_3O_4 layer at the electrode surface (see Eq. (6.1)). In the case of an aged battery, an amount of electrochemically active Li^+ ions will remain stored in the LiC_6 electrode, resulting in a $Q_{a\,max}$ higher than $Q^+_{a\,max}$. The formation of the Co_3O_4 at the $LiCoO_2$ electrode surface will also contribute to the increase in battery overpotential. During a battery's lifetime, the amount of Li^+ ions remaining inside the graphite at the end of discharging under the same operational conditions will therefore increase. This is represented by Q_0^- and Q_{a0}^- in Fig. 6.23. It should be noted that during a battery's lifetime an amount of the Li^+ ions will also be consumed in the Solid Electrolyte Interface (SEI), in an irreversible process [1].

Given the parameters shown in Fig. 6.23 and the experimentally observed SoC values, E^+_{eq} and E^-_{eq} of an aged battery can be determined in the same way using the model presented in chapter 4. From the situation described above it follows that the EMF model parameters relating to battery design will change during a battery's lifetime. This means that, to obtain accurate SoC determination, the changing of these parameters with a battery's aging must be taken into account.

Fig. 6.24 illustrates the E^+_{eq} and E^-_{eq} dependence on SoC in the case of a fresh and an aged battery.

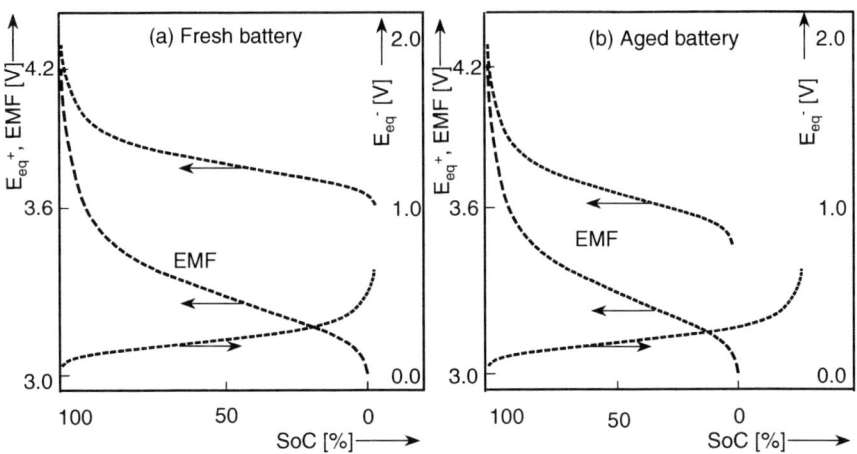

Fig. 6.24. Schematic representation of the EMFs and positive and negative electrode EMF voltages of a fresh (a) and an aged (b) battery.

E^+_{eq} may have a voltage value between 3.6 and 4.3 V, whereas E^-_{eq} ranges between 0.1 and 0.6 V. The EMF calculated for the fresh battery consequently has a voltage value between 3 and 4.2 V. These EMF limits correspond to 0 and 100% SoC, respectively. During the battery's lifetime the $LiCoO_2$ electrode will decompose and an amount of the cyclable Li^+ ions will remain stored inside the LiC_6 electrode. Therefore, $Q^+_{a\,max}/2$ will be lower than $Q^-_{a\,max}$. E^-_{eq} of an aged battery will consequently have a EMF value lower than 0.6 V at the 0% SoC level (see Fig. 6.24 (b)). To reach the defined 3 V EMF level at 0% SoC, the positive electrode must consequently be discharged to below the 3.6 V EMF level (see Fig. 6.24 (b)). The EMF of the aged battery will then be calculated as a difference between different equilibrium potential levels with respect to the EMF calculated for a fresh battery. The EMF obtained for an aged battery will therefore differ from the EMF obtained for a fresh battery. In addition, the process of the battery's degradation, induced mainly by the $LiCoO_2$ electrode, will be accelerated.

The battery model presented in [13] was simulated in order to further investigate the EMF dependence on the aging effect. The aspects presented above were taken into account during the simulations. Good agreement was obtained between the qualitative explanation given in Fig. 6.24 and the results of the simulations. Furthermore, several batteries have been opened and measurements with reference electrodes have been performed. These measurements confirmed that the positive electrode decomposition mechanism is mainly responsible for the EMF aging effect. This enhances the confidence level of the EMF dependence on the aging process presented in this chapter.

6.3 Overpotential dependence on battery aging

New overpotential measurements and modelling results obtained for Li-ion batteries as a function of aging will be presented in this section. The overpotential measurement results obtained for a fresh battery on the basis of partial charge/discharge cycles will be compared with those obtained for aged batteries. The temperature influence will also be considered in this comparison. The voltage-relaxation model presented in sections 4.2 and 6.2 will be used to study the influence of a not-fully-relaxed voltage in the battery overpotential determination. New results regarding the overpotential symmetry as a function of battery aging will also be presented.

6.3.1 Overpotential measurements as a function of aging

As the overpotential represents the difference between the EMF and the charge/discharge voltage, a charge/discharge EMF was first experimentally determined by means of GITT as described in section 6.2. During this measurement method the battery was charged/discharged in identical steps of 4% SoC at 0.1 C-rate current. Each charge/discharge step was followed by a rest period chosen as a function of SoC. In order to obtain information on the battery overpotential, the difference between the EMF measured at the end of the relaxation period and the voltage last measured after a charge/discharge step was considered. This led to a charge overpotential with a negative value and a discharge overpotential with a positive value.

A major advantage of using GITT voltage-relaxation measurements for aged battery overpotential calculation is that interpolation calculations are not needed for the charge/discharge battery overpotential comparison. The difference between the EMF and the voltage last measured after a charge/discharge step is calculated at each 4% SoC. This eliminates the inaccuracy that may result from the interpolation of the overpotential measurements from the charge/discharge battery overpotentials comparison.

Examples of the relaxation processes in a fresh and an aged battery are illustrated in Fig. 6.2. It follows from Fig. 6.2 and the situation described above that the overpotential increases with battery aging, which agrees with the aspects of battery aging considered in section 6.1. Fig. 6.25 shows the battery overpotentials calculated for a fresh and an aged Li-ion battery after charge and discharge steps at 0.1 C-rate current and 25°C. It can be concluded from this figure that the maximum overpotential is 224 mV in the case of the 25.4% capacity loss battery at 2.9% SoC.

The GITT voltage-relaxation measurements showed that, especially at low SoC values, a 24 hours' rest is not always sufficient for a battery to reach the equilibrium voltage, *i.e.* EMF. In order to investigate the influence of a not-fully-relaxed voltage in the battery overpotential calculation, the battery voltage was further predicted by using the new voltage-relaxation model presented in section 4.2. The voltages measured during the rest periods were used as input for the voltage-relaxation model. In this example a rest period of 96 hours was for practical reasons considered for the voltage prediction in the fresh and aged batteries. Fig. 6.26 shows the battery charge/discharge overpotentials calculated as a difference between the predicted battery voltage after 96 hours and the voltage last measured after a charge/discharge step.

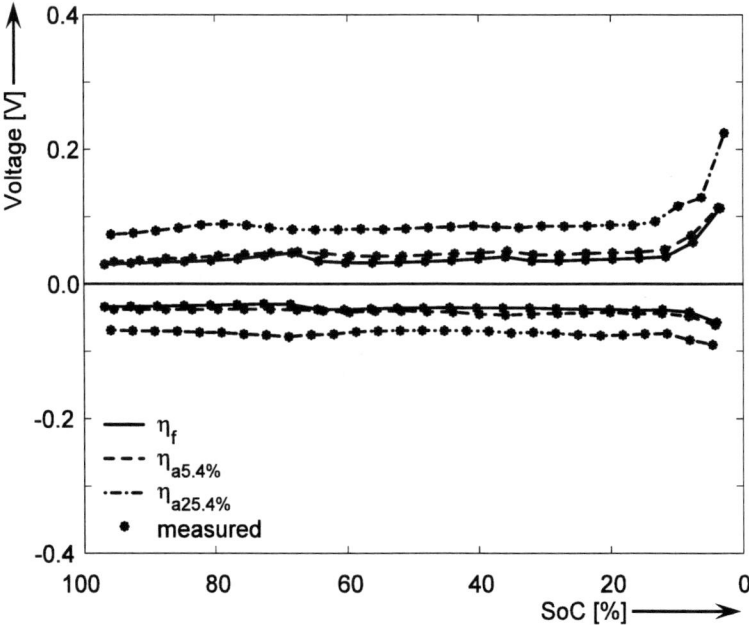

Fig. 6.25. The overpotentials of fresh (*f*) and aged (*a*) batteries measured at 25°C.

The charge overpotential determined for fresh and aged batteries as a difference between EMF and V_{chl} and the charge overpotential determined as a difference between V_p and V_{chl} are consistently the same at all SoC values. Another conclusion relates to the discharge overpotential. The discharge overpotential determined for the fresh and aged batteries as a difference between EMF and the voltage last measured after a discharge (V_{dl}) step and the discharge overpotential determined as a difference between V_p and V_{dl} are consistently the same at SoC values higher than 6.7%. Only a small voltage difference is observed at low SoCs between the discharge overpotentials obtained using the methods considered above. It can be concluded that 24 hours' rest at low SoC values and 12 hours' rest at high SoC values will usually be sufficient to ensure an accurate battery overpotential in fresh and aged batteries at 25°C. It follows from Fig. 6.26 that the maximum overpotential of the 25.4% capacity loss battery is 244 mV after a discharge step at 2.9% SoC (see section 6.1).

From the measurements based on partial charge/discharge steps it follows that the battery overpotential increases at low SoC values and remains almost constant at SoC values higher than 20%. Another important conclusion relates to the shape of the battery charge/discharge overpotentials. From Figs. 6.25 and 6.26 it follows that the mean calculated charge/discharge battery overpotential of fresh and aged batteries is symmetrical between 20 and 80% SoC.

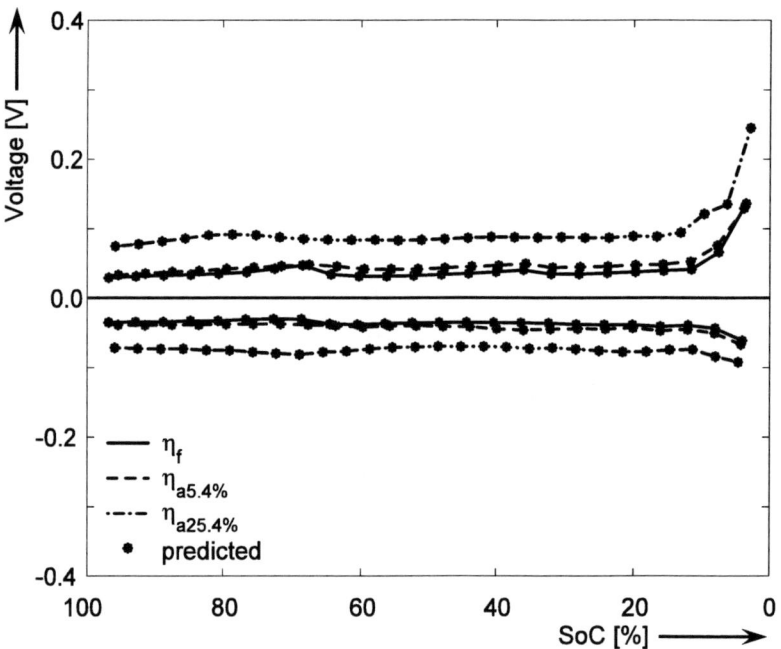

Fig. 6.26. The overpotentials of fresh (*f*) and aged (*a*) batteries measured at 25°C taking into account the voltage-relaxation model.

The partial charge/discharge battery overpotential measurements in fresh and aged batteries as a function of temperature will be further discussed in this section. In order obtain information on the battery overpotential dependence on temperature, the GITT and voltage-relaxation methods described above were applied at 5 and 45°C. Fig. 6.27 shows the battery overpotential calculated after charge and discharge steps of 4% SoC, at 0.1 C-rate current and 5°C. The maximum overpotential of the 25.4% capacity loss battery was evidently 310 mV after a discharge step at 3.2% SoC.

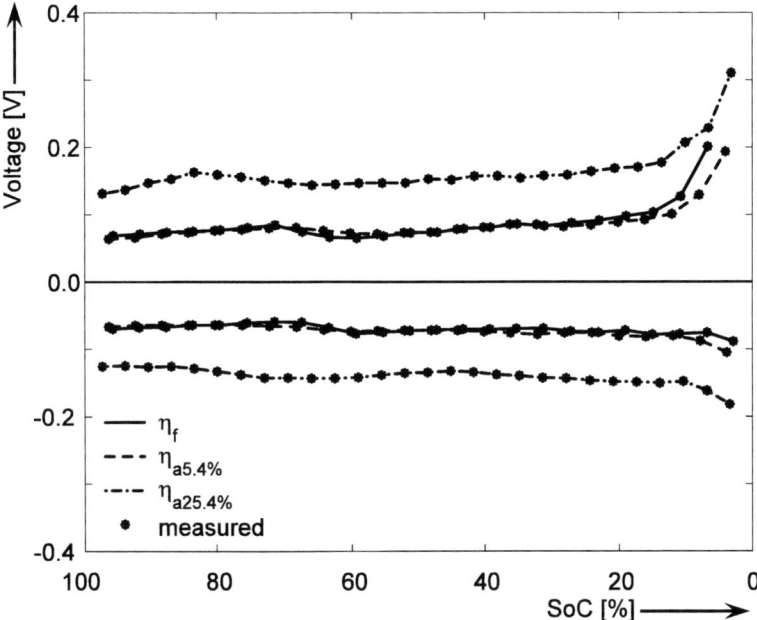

Fig. 6.27. The overpotentials of fresh (*f*) and aged (*a*) batteries measured at 5°C.

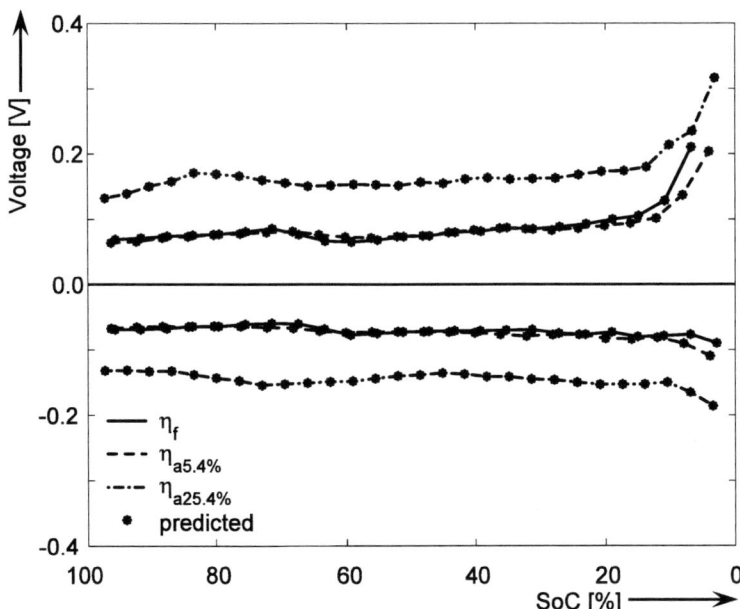

Fig. 6.28. The overpotentials of fresh (*f*) and aged (*a*) batteries measured at 5°C with the voltage-relaxation model plotted as a function of SoC [%].

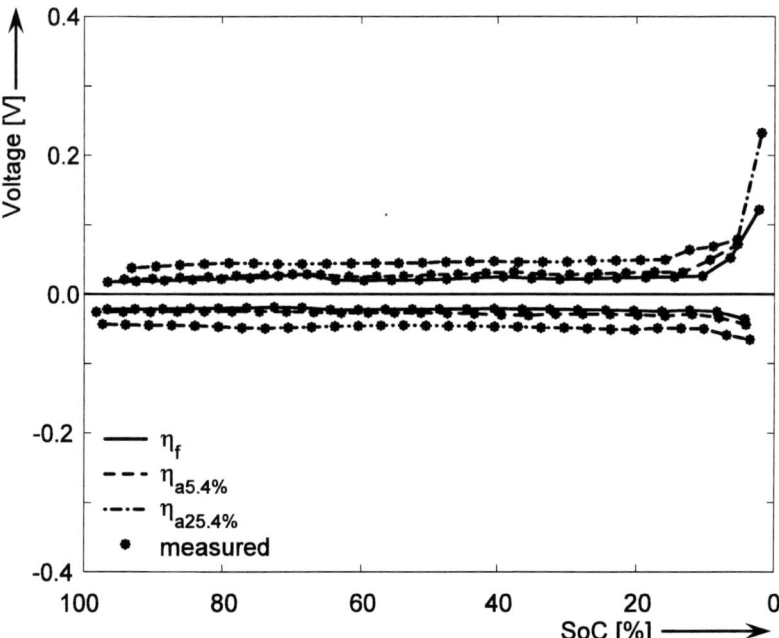

Fig. 6.29. The overpotentials of fresh (*f*) and aged (*a*) batteries measured at 45°C.

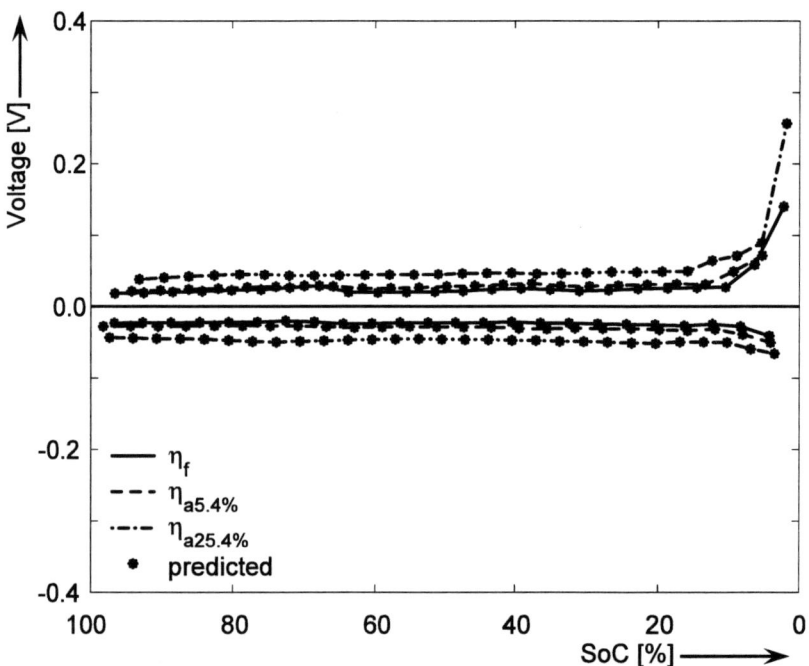

Fig. 6.30. The overpotentials of fresh (*f*) and aged (*a*) batteries measured at 45°C with the voltage-relaxation model plotted as a function of SoC [%].

To enable comparison, the battery voltage was further predicted by using the new voltage-relaxation model presented in section 4.2. Fig. 6.28 shows the battery charge/discharge overpotentials calculated as a difference between the predicted battery voltage after 96 hours and the voltage last measured after a charge/discharge step. A maximum overpotential of 317 mV was obtained for the 25.4% aged battery after a discharge step at 3.2% SoC.

Fig. 6.29 shows the battery overpotential after charge and discharge at 45°C. The maximum overpotential of the 25.4% capacity loss battery is 231 mV after a discharge step at 1.9% SoC. When the voltage prediction model is also considered the maximum overpotential increases by a value of 256 mV. This situation is presented in Fig. 6.30.

The results presented in Figs. 6.27–6.30 are again similar to the results obtained at 25°C, confirming that the rest periods were long enough for an accurate overpotential calculation and that the symmetry of charge and discharge overpotentials is also observed at other temperatures. The measurements based on partial charge/discharge steps show that the battery overpotential increases with aging at low SoCs and temperatures.

6.4 Adaptive systems

In order to enable accurate battery modelling, new adaptive systems based on phenomena discovered during the analysis of the measurements of the battery Electro-Motive Force and overpotential models will be developed in this section. These adaptive systems are necessary because, as shown in sections 6.2 and 6.3, the behaviour of a battery's EMF and overpotential changes with time due to the aging processes. It should be added that EMF and overpotential model parameter values are determined for a targeted battery chemistry/type of battery. Due to manufacturing spread not all batteries from the same manufacturing batch will behave identically. What's worse is that the behaviour of a battery's EMF and overpotential will change with the battery's chemistry [1]. The parameter values must therefore be updated to ensure accurate modelling of batteries with different chemistries during their life cycles.

6.4.1 Electro-Motive Force adaptive system

The proposed EMF adaptive method is based on the maximum capacity and GITT measurement methods (see Section 6.2) combined with the newly developed voltage-relaxation model (see section 4.2). In this chapter, the adaptive EMF method will be considered by applying the following measurement method. According to this method, a battery's maximum capacity is determined during a complete charge cycle from a low SoC value, *i.e.* lower than 1% SoC. The battery is charged using the normal CCCV charging method [1]. The maximum capacity is calculated using the method described in [14]. The charging step is followed by a rest period of about 4 hours. In our calculation, the battery EMF voltage after the rest period had a value of 4.175 V and the SoC level was defined to be 100%. The battery was further discharged from 4.175 V at 0.1 C-rate current in steps of 4% SoC. The discharge step was followed by a rest period of 12 hours. By the end of the rest period the battery had reached the equilibrium state. In this way a first EMF point, EMF_1, with the corresponding SoC, was determined (see Fig. 6.31). The discharging was repeated until the battery voltage reached 3 V. The measurements were performed at 5, 25 and 45°C. An example measurement carried out at 7 EMF points using discharge steps of 4% SoC at 25°C is illustrated in Fig. 6. 31.

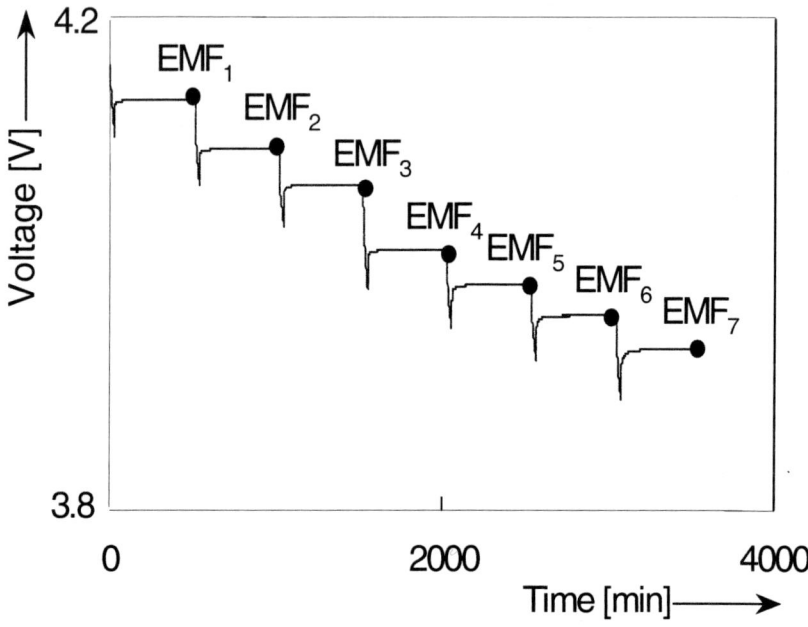

Fig. 6.31. The EMF points (EMF_1, ..., EMF_7) measured after discharge steps of 4% SoC at 0.1 C-rate current and 25°C.

The chosen C-rate current, SoC step and rest period make the EMF adaptation method easy to implement, but the method described above is not restricted to any specific C-rate current, discharge SoC step or rest period, and can consequently be used for varying conditions. The newly developed voltage-relaxation model could for example be used to avoid the long rest periods. Analysis of the voltage-relaxation results showed that a rest period of 15 minutes rest will always lead to SoC results that are accurate to within 0.5% SoC under a broad range of conditions. For this reason, a rest period of 15 minutes can be considered sufficient for accurate determination of a battery's EMF.

An example of EMF adaptation in the 5.4% capacity loss battery (see section 6.2) at 25°C will be further considered. In this example, 10 EMF predicted points are considered during discharging. The EMF predicted points are distributed along the horizontal axis. The voltage and time samples measured during the first 15 minutes of the rest period are considered in this prediction. The 0 and 100% SoC levels with the corresponding EMF values, i.e. 3 and 4.175 V, respectively, are also considered in this example. The 12 EMF points are further fitted using the newly developed method in which the shape of the curve is also taken into account. The calculated EMF is illustrated in Fig. 6.32. To enable comparison, the EMF measured by means of GITT is also shown. The difference between the EMFs obtained by means of GITT and by means of the method described above is shown in Fig. 6.33.

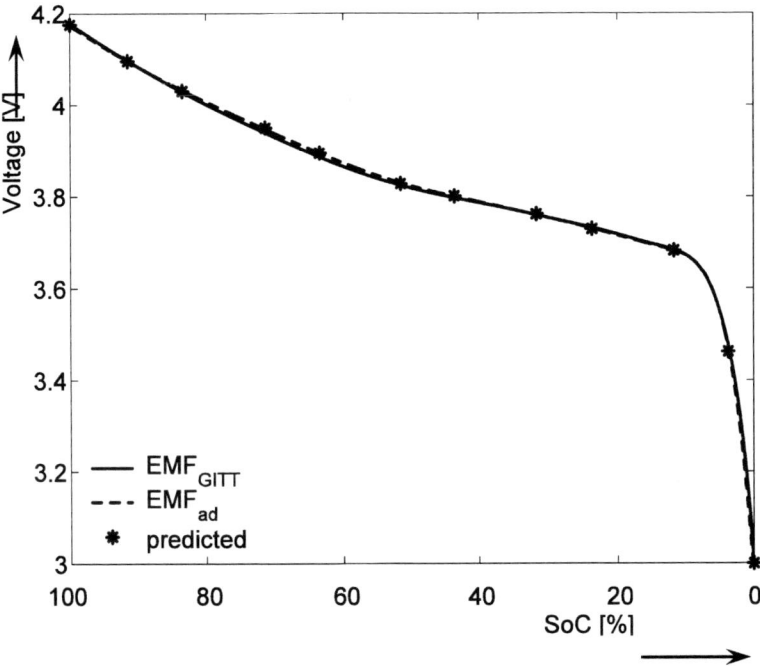

Fig. 6.32. EMF obtained by means of 12 EMF predicted points (*) and the described adaptation method (EMF_{ad}) compared with EMF obtained by means of GITT (EMF_{GITT}) at 25°C using a 5.4% capacity loss battery.

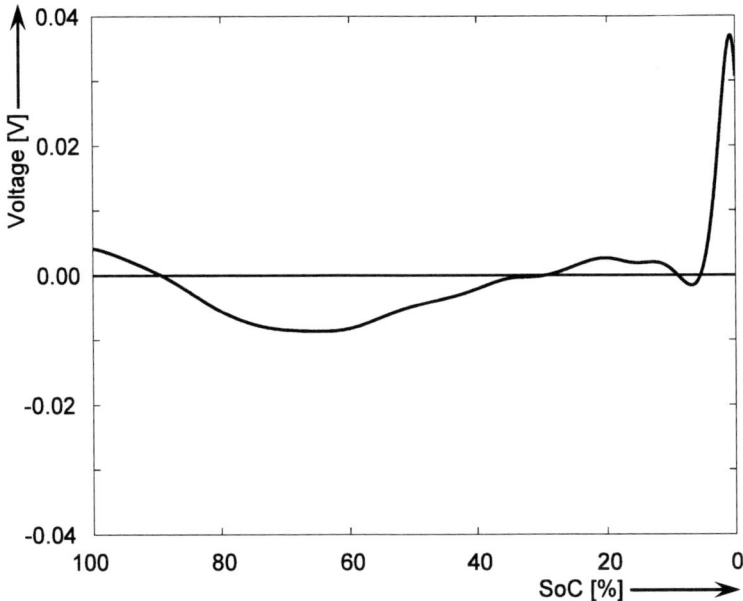

Fig. 6.33. The difference between EMF obtained by means of GITT and EMF obtained by means of the described adaptation method (EMF_{ad}) at 25°C (see also Fig. 6.32).

It can be seen in Figs. 6.32 and 6.33 that a maximum difference in EMF of 36 mV was obtained at 1.1% SoC. This means that when the EMF adaptation method is used the SoC indication system based on EMF will in this case yield an SoC value of 1.3% whereas the actual SoC value calculated on the basis of the discharge EMF is 1.1%. The inaccuracy will be –0.2% SoC. This effect will be more pronounced in the flat region of the EMF-SoC curve, where even minor differences in EMF will cause substantial errors in SoC. A difference of about 8 mV will for example be obtained between the EMF obtained using GITT and that obtained with the described adaptation method at 67% SoC. This means that when the EMF adaptation method is used, the SoC indication system based on EMF will in this case yield an SoC value of 66% whereas the actual SoC value calculated on the basis of the discharge EMF is 67%. The inaccuracy will be 1% SoC. It can be concluded that the newly developed EMF adaptation method will always yield an SoC that is accurate to within 1%. The EMF adaptation accuracy can be improved by considering a longer relaxation time in the voltage-relaxation model or more EMF points in the fitting method.

6.4.2 Overpotential adaptive system

Besides a battery's EMF curve, its overpotential development will also change over time. When this change in overpotential behaviour is not taken into account the accuracy of the overpotential model described in chapter 5 will gradually decrease as a battery ages.

The main problem with overpotentials is that they cannot be measured directly (see section 5.1). When a battery's voltage has been measured and its EMF is known, an estimate of the overpotential can however be derived. A remaining difficulty is the fact that the overpotential depends on many factors, including a battery's SoC, current, temperature, time, spread and age. Updating should therefore be initiated when most of these variables are constant. This section presents a new method for adapting a battery's overpotential in accordance with the aging effect. The symmetry in the overpotential will also be used in this adaptive system.

The mechanism proposed for updating the overpotential model parameters takes advantage of the fact that the updating is performed during charging. The current will then be constant in CC mode and the temperature may also be considered constant because the charger will usually be used in-house, where temperature variations are limited. During normal CC charging the charge current will moreover not be interrupted, so after the overpotentials have been built up in the initial stages of charging the time variable will not play a dominant role.

The principle of the overpotential adaptive method is as follows. The overpotential model parameters of a fresh battery (η^f) are obtained by fitting the overpotential model (see Eq. (5.1)) to measured discharge overpotentials (η_d^f) obtained at four C-rate currents together with the measured charge overpotential (η_{ch}^f) obtained during CC charge mode (see chapter 5). η_{ch} can also be measured for each SoC during CC charge mode by subtracting the EMF value from the measured battery voltage value. The SoC expressed as a percentage can be used to assess the EMF value during CC charge mode using the SoC-EMF relationship described in chapter 4. During a battery's lifetime the charge overpotential (η_{ch}^a)

can therefore be continuously measured by using the η_{ch} measurement method described above. A ratio at the same SoC, C-rate current and temperature can consequently be calculated for the fresh and the aged battery charge overpotentials. As shown in this chapter, the charge and discharge overpotentials are symmetrical with respect to the horizontal axis. For this reason, the $\eta_{ch}^a / \eta_{ch}^f$ ratio determined for the charge battery overpotential also works well for the discharge battery overpotential. The discharge overpotential of an aged battery (η_d^a) can consequently be determined by using Eq. (6.5).

$$\eta_d^a = \eta_d^f \frac{\eta_{ch}^a}{\eta_{ch}^f} \qquad (6.5)$$

It should be noted that the charge and discharge overpotentials of a fresh and an aged battery must be determined at the same SoC, C-rate current and temperature.

To enable accurate measurement of the charge overpotential during a battery's lifetime, the EMF model parameters should also be adapted using the adaptation method described earlier in this section.

6.5 Conclusions

A description of the behaviour of a battery's EMF and overpotential as a function of aging has been presented. The measurement and modelling methods presented in this chapter have been applied to fresh and aged US18500G3 Li-ion batteries (Sony). As shown above, battery aging is a complex process that involves many battery parameters, e.g. impedance and capacity, the latter being the most important. These two factors must be correctly identified and separately analysed to obtain accurate battery measurements and modelling methods.

The EMF results obtained for two aged batteries by using GITT have been compared with those obtained for fresh batteries. Accurate results were obtained for the aged batteries, too, by applying the voltage-relaxation model under a broad range of conditions. This proved to be important for an accurate interpretation of EMF measurements obtained for aged batteries. The effects of temperature and hysteresis on the EMF curves of aged batteries have been discussed. These measurements show that the EMF depends on the aging effect. For this reason, the assumption regarding EMF presented in [1]–[5], i.e. that the EMF of a Li-ion battery depends on aging to only a limited extent, does not hold for US18500G3 Li-ion batteries. The EMF mathematical model has been further used to explain the EMF dependence on battery aging. The difference in EMF is assumed to be mainly attributable to the decomposition of the $LiCoO_2$ electrode.

Two possible methods for battery overpotential determination as a function of the aging effect have been presented. In the first method the charge/discharge overpotential is obtained via partial charge/discharge steps at different temperatures. The EMF measured by means of the Galvanostatic Intermittent Titration Technique (GITT) and the voltage-relaxation model have also been used in this method. The advantage of using the method based on partial charge/discharge steps is that the GITT measurements can be used directly to obtain information on the charge/discharge battery overpotential.

The main drawback of the method based on partial charge/discharge steps is that full information on the build-up of an aged battery's overpotential in time cannot be obtained from the applied charge/discharge period. This information was obtained by subjecting aged batteries to new measurements using a second method. The employed two measurement methods show that the value of a battery's overpotential depends on the charge/discharge C-rate current, charge/discharge period, SoC, temperature and aging. A battery's overpotential will for example increase at low SoCs and temperatures and with aging. The overpotential symmetry phenomenon discovered during the measurements has been discussed in relation to aged batteries. It has been shown that the calculated charge/discharge battery overpotential is symmetrical between 20 and 80% SoC. Aspects of this phenomenon have been used to adapt the overpotential model with aging.

This chapter also presents new adaptive systems for measuring a battery's Electro-Motive Force and overpotential models for enabling accurate SoC determination. The EMF and overpotential model parameter values need to be updated in order to enable accurate modelling of batteries with different chemistries during the batteries' life cycle. The EMF adaptive method combines the maximum capacity and the GITT measurement methods with the newly developed voltage-relaxation model. A possible example of EMF adaptation for a 5.4% capacity loss battery by means of 10 EMF predicted points has been considered. This prediction is based on voltage and time samples measured during the first 15 minutes of the rest period. In this example the maximum inaccuracy in SoC prediction was found to be 1%. The basis of the proposed overpotential adaptive method is the overpotential symmetry. The mechanisms proposed for updating the overpotential model also take advantage of the fact that the update is performed during charging. A major advantage is that external conditions during a battery's charging, such as charge current and battery temperature, will be constant.

The assumptions made in the considered EMF and overpotential adaptation methods make the update mechanism easy to implement, but the described methods are not restricted to any specific current or temperature, and may therefore be used for varying conditions. The presented adaptive systems should ultimately update EMF and overpotential model parameters and expand the model's potential to include other types of batteries, too. More information on implementation aspects and test results obtained for aged batteries and different types of batteries will be given in chapter 8.

6.6 References

[1] H.J. Bergveld, W.S. Kruijt, P.H.L. Notten, Battery Management Systems, Design by Modelling, Philips Research Book Series, 1, Kluwer Academic Publishers, Boston (2002)

[2] F.A.C.M. Schoofs, W.S. Kruijt, R.E.F. Einerhand, S.A.C. Hanneman, H.J. Bergveld, Method of and device for determining the charge condition of a battery, US 6,420,851, filed April 4 (2000)

[3] F.A.C.M. Schoofs, W.S. Kruijt, R.E.F. Einerhand, S.A.C. Hanneman, H.J. Bergveld, Method of and Device for Determining the Charge Condition of a Battery, US Patent 6,420,851, filed March 29 (2000)

[4] V. Pop, H.J. Bergveld, P.H.L. Notten, P.P.L. Regtien, State-of-Charge Indication in Portable Applications, IEEE International Symposium on Industrial Electronics, **3**, 1007–1012 (2005)
[5] Texas Instruments, Portable Power Design Seminar (2004)
[6] J.H. Aylor, A. Thieme, B.W. Johnson, A battery State-of-Charge Indicator for Electric Wheelchairs, IEEE Trans. On industrial electronics, **39**, 398–409 (1992)
[7] T. Kikuoka, H. Yamamoto, N. Sasaki, K. Wakui, K. Murakami, K. Ohnishi, G. Kawamura, H. Noguchi, F. Ukigaya, System for measuring state of charge of storage battery, US Patent 4,377,787, filed August 8 (1980)
[8] G.R. Seyfang, Battery state of charge indicator, US Patent 4,949,046, filed June 21 (1988)
[9] M.W. Verbrugge, E.D. Tate Jr., S.D. Sarbacker, B.J. Koch, Quasi-adaptive method for determining a battery's state of charge, US Patent 6,359,419, filed December 27 (2000)
[10] T. Zheng, W.R. McKinnon, J.R. Dahn, Hysteresis During Lithium Insertion in Hydrogen-Containing Carbons, J. Power Sources, **143**, 2137–2145 (1996)
[11] T. Zheng, J.R. Dahn, Hysteresis Observed in Quasi-open Circuit Voltage Measurements of Lithium Insertion in Hydrogen-containing Carbons, J. Power Sources, **68**, 201–203 (1997)
[12] V. Srinivasan, J.W. Weidner, J. Newman, Hysteresis during Cycling of Nickel Hydroxide Active Material, J. of Electrochem. Soc., **148**, 969–980 (2001)
[13] D. Danilov, P.H.L. Notten, 12th International Meeting on Lithium Batteries, Nara (2004)
[14] H. J. Bergveld, V. Pop and P.H.L. Notten, Method of estimating the State-of-Charge and of the use time left of a rechargeable battery, and apparatus for executing such a method, Patent WO2005085889, filed February 23 (2005)

Chapter 7
Measurement results obtained with new SoC algorithms using fresh batteries

A complete description of EMF and overpotential behaviour of US18500G3 Li-ion batteries has been given in chapters 4 and 5 of this book. In this chapter, a new SoC algorithm inferred from the SoC system described in [1] will be developed. The algorithm combines adaptive and predictive systems with Electro-Motive Force (EMF) measurement during the equilibrium state and Coulomb counting (cc) [1]–[12] during the charge and discharge states. Besides cc, the effect of the battery overpotential (η) during discharge will be also considered. The goal of the SoC system is to predict the remaining run-time (t_r) of an Li-ion battery with an uncertainty of 1 minute or less under all realistic user conditions, including a wide variety of load currents and a wide temperature range.

Basic issues concerning SoC and t_r are presented in section 7.1. A new SoC algorithm and the method of implementing the algorithm states in a real-time SoC evaluation system are presented in section 7.2. The focus in section 7.3 is on the experimental results obtained with the real-time SoC evaluation system. The identification of sources of error in each state of the real-time SoC evaluation system is discussed in section 7.4. The conclusions of the error sources analysis are used to develop an improved SoC algorithm in section 7.5. They are also compared with the SoC algorithm introduced in section 7.2. Section 7.6 presents a comparison with a competitive SoC indication system, i.e. the bq26500 developed by Texas Instruments. These results prove that it is indeed possible to test SoC-indication algorithms with the newly developed SoC evaluation system and show that the new approach is effective in improving remaining run-time indication accuracy. Finally, section 7.7 presents concluding remarks.

7.1 Introduction

As already mentioned in the previous chapters of this book, several methods for SoC calculation are available in practice. This chapter presents a new method for predicting a battery's SoC that aims to eliminate the main drawbacks and combine the advantages of the battery-related models described in chapters 4 and 5, respectively. The advantages of Electro-Motive Force (EMF) and overpotential will be combined in a new SoC indication algorithm. In order to ensure accurate SoC indication, the voltage-relaxation predictive method [13] and a new maximum capacity adaptive method [10] will be also considered in the SoC algorithm. The SoC calculation and battery overpotential prediction will also be used to calculate the remaining run-time available under the valid discharge conditions. In order to prove SoC and remaining run-time accuracy, the new SoC algorithm will be further implemented in a real-time SoC evaluation system.

7.2 Implementation aspects of the algorithm

7.2.1 A new SoC algorithm

The basic structure of the SoC algorithm is schematically represented in Fig. 7.1.

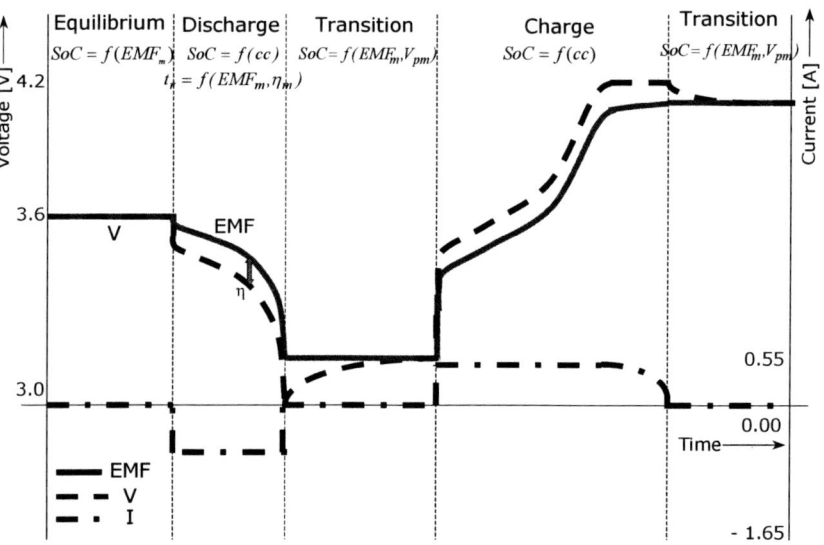

Fig. 7.1 Schematic representation of the SoC algorithm states. The horizontal axis shows the time.

The SoC algorithm starts up in the initial state (not represented in Fig. 7.1). In this state the initial SoC is determined on the basis of voltage (V) and temperature (T) measurements and the EMF model (EMF_m) developed in section 4.4. Dependent on whether the battery is being charged or discharged or is in equilibrium, the algorithm then shifts to the appropriate state.

In the equilibrium state the battery voltage (V) is stable and SoC is determined by means of voltage and temperature measurements and the EMF model (see Fig. 7.1).

In the discharge state SoC is determined by means of Coulomb counting, expressed as cc in Fig. 7.1. In addition to simple cc the effect of the overpotential is also considered (see Fig. 7.1). As will be shown later in this section the prediction of the overpotential also yields a remaining run-time prediction. Results of our mathematical overpotential implementation were presented in section 5.2.

In the charge state SoC is determined via Coulomb counting (see Fig. 7.1). The stable conditions of the charge state will be used in the SoC evaluation system in order to adapt the system to make allowance for the aging effect. One of the main advantages, which is also independent of the type of charge method used, is that the environmental temperature will in most practical cases be constant during the charge state.

The transitional state is used when the algorithm changes from either the charge or the discharge state to the equilibrium state. In this state it is determined whether the battery voltage is stable and the algorithm is allowed to enter the equilibrium state. SoC is in this state determined by means of the voltage-prediction model (V_{pm}) and EMF_m (see Fig. 7.1). Results of V_{pm} were presented in section 4.2. The algorithm is assumed to be in the equilibrium state when the difference between the voltage predicted by means of the voltage-prediction model and the measured battery voltage is smaller than 1 mV.

In summary, in which state the algorithm is operating will depend on the value and sign of the current flowing into or out of the battery and on whether the battery voltage is stable or not, *i.e.* whether it has reached an EMF value.

In order to check the SoC and the remaining run-time accuracy, the new SoC algorithm was implemented in a real-time SoC evaluation system that operates in initial, standby, backlight-on, transitional, charge and discharge states. A state diagram illustrating the basic structure of the real-time SoC evaluation system is shown in Fig. 7.2. The backlight-on state is not included in this figure.

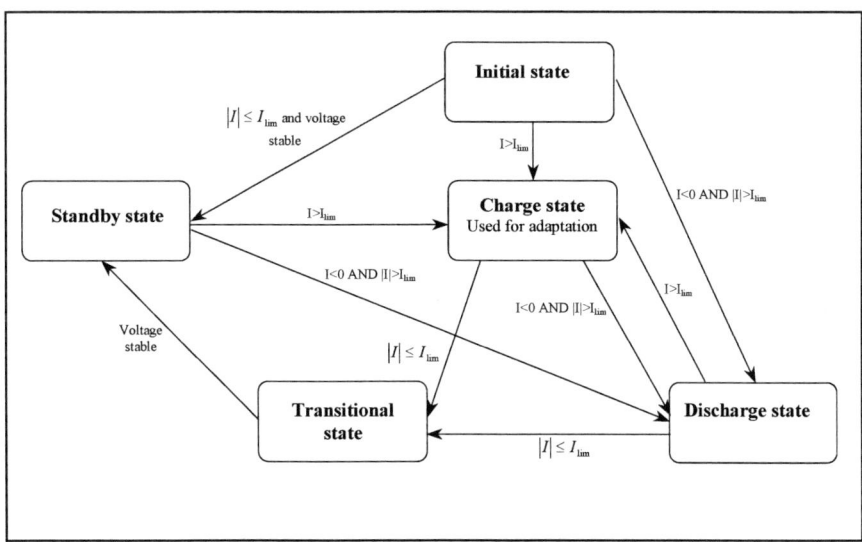

Fig. 7.2. State diagram of the real-time SoC evaluation system.

Each time the SoC evaluation system is switched on, it will start in the initial state (see Fig. 7.2). The SoC is in this state determined in the same way as in the initial state of the SoC algorithm.

In the standby state hardly any current is drawn from the battery. In this situation the battery is in equilibrium after full relaxation (see Fig. 7.1). The SoC is in the standby state (SoC_s) therefore determined by means of V and T measurements and the stored EMF_m. The current in the standby state is only a few mA, *i.e.* 1 mA in the SoC evaluation system described in this section, which is lower than the 10 mA current limit (I_{lim}) defined in the SoC evaluation system (see Fig. 7.2). At this very low standby current value, the battery voltage will be very close to the EMF value, providing the voltage is stable. So, in order to allow the SoC evaluation system to change to this state, the condition of stable voltage has to be met. From this state the system is able to switch to the charge, discharge, or backlight-on states.

In the backlight-on state a small negative current, *i.e.* 6 mA in the SoC evaluation system described in this section, is drawn from the battery. This will be the case when a user activates the screen of the SoC evaluation system. Because the current is still below the I_{lim} value, SoC will still be determined by means of V and T measurements and the stored EMF_m. The SoC evaluation system will remain in this state for about 5 seconds (an arbitrarily chosen value) before automatically switching to the standby state. For these 5 seconds, all other transitions to the charge or discharge states will remain possible.

In the transitional state a small negative current, *i.e.* 1 mA in the SoC evaluation system described in this section, is drawn from the battery. The SoC is in this state determined in the same way as in the transitional state of the SoC algorithm.

In the charge state, a charger is connected to the battery and a positive current larger than the I_{lim} value flows into the battery (see Fig. 7.2). SoC in the charge state is determined by means of cc. The stable conditions during the charge state are exploited to adapt the maximum capacity value (Q_{max}) to allow for the aging effect.

As shown in chapter 6, any battery will lose capacity during cycling (see Fig. 6.1). In order to ensure accurate SoC and t_r calculation and to improve the SoC evaluation system capability to allow for the aging effect, a simple method for updating the maximum capacity taking capacity loss into account will be presented. In this method the stable conditions during charging are exploited to adapt Q_{max} to allow for the aging effect [10]. Fig. 7.3 shows how the maximum capacity is updated.

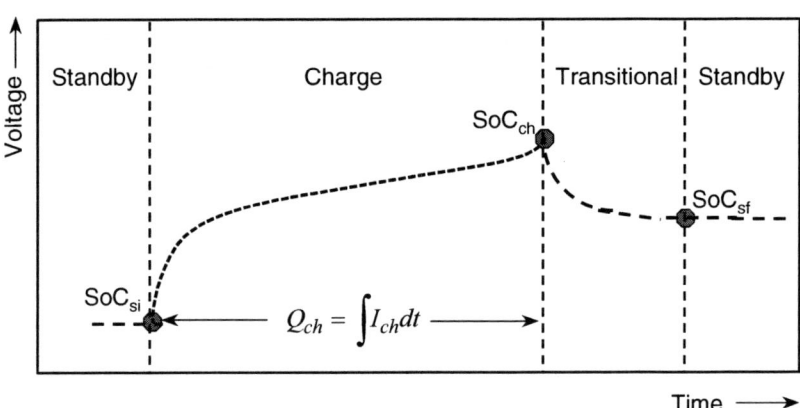

Fig. 7.3. Schematic representation of the method for updating Q_{max} taking capacity loss into account.

It is necessary for the system to run through a sequence of states: standby state, charge state, transitional state and standby state (see Fig. 7.3). A new value of Q_{max} is determined as follows

$$Q_{max} = \frac{100}{SoC_{sf} - SoC_{si}} Q_{ch} \tag{7.1}$$

where SoC_{si} and SoC_{sf} denote the initial and final State-of-Charge in the standby state in [%] and Q_{ch} denotes the amount of charge flowing into the battery during the charge state in [mAh].

The greatest advantages of this algorithm over the maximum capacity adaptation algorithm presented in [1] (see also Fig. 3.6) are:

- during the charge state the current and temperature are more or less constant. These two variables can therefore be measured more easily than in the discharge state;
- in a practical application the charging period will usually be longer than the discharging period, as a result of which, the Q_{max} inaccuracy caused by the inaccuracy generated in the SoC_{sf} and SoC_{si} determination will be smaller (see also section 7.4);
- during the CV mode the overpotential will be smaller (see chapter 5), so the algorithm will reach the standby state sooner. This will improve the SoC_{sf} determination accuracy;
- if a battery is charged until the defined end-of-charge conditions are reached, *i.e.* 0.05 C-rate current in the CCCV algorithm, the SoC_{sf} will automatically be 100%. The inaccuracy introduced at this point in the Q_{max} determination will consequently be very small. More information on the maximum capacity determination accuracy will be given in section 7.4.

In the discharge state, a negative current I_d larger in modulus than the I_{lim} value flows out of the battery (see Fig. 7.2). SoC in the discharge state, SoC_d, is determined by means of cc. In addition to SoC_d, the remaining run-time available under the discharge conditions is also calculated during the discharge state. So besides simple cc, the effect of the battery overpotential is also considered in the remaining run-time calculation. The predicted battery overpotential at the 3 V End-of-Discharge Voltage (V_{EoD}) value is translated into a SoC-left percentage value SoC_l by using the EMF model (see Fig. 3.4: SoC_l = SoC at point B). t_r [min.] is consequently inferred from SoC_d [%], SoC_l [%], Q_{max} [mAh] and I_d [A] according to

$$t_r [\min] = \frac{0.06 \frac{Q_{max}}{100}(SoC_d - SoC_l)}{I_d} \quad (7.2)$$

At the end of the discharge state, the system passes through the transitional state to the standby state. It can be concluded from the implementation method described above that in the charge and discharge states SoC is calculated by means of cc. Besides cc, the overpotential is also predicted for the t_r calculation in the discharge state.

In summary, in which state the real-time SoC evaluation system is operating will depend on the value and sign of the current flowing into or out of the battery and on whether the battery voltage is stable or not.

7.2.2 Implementation aspects of the SoC algorithm

A performance analysis of the real-time SoC evaluation system must be carried out to test the accuracy of the SoC algorithm. This section describes the hardware structure of the real-time evaluation system. A test set-up was designed containing a computer with a National Instruments (NI) Data-Acquisition interface card, an SCB-68 National Instruments connector board, a 20 mΩ sense resistor (R_S,) a Keithley 2420 3A Source Meter device with several digital Input/Output pins (Dig. I/O) used for (dis)charging the battery, a temperature box that can keep the battery at a constant temperature between $-35°C$ and $65°C$, an LM335 National Semiconductor precision temperature sensor with an accuracy within $\pm 1°C$ connected to the SCB-68 board and a safety box used as a device to prevent over(dis)charging of the battery. The SCB-68 board is a shielded board with 68 screw terminals for easy connection to NI 68-pin products. The wiring diagram of the real-time operation system is given in Fig. 7.4. The PC and temperature box are not shown in this figure.

The battery voltage, current and temperature have to be monitored and the safety box has to be controlled to ensure that the Li-ion battery is never operated in an unsafe region. The temperature-sensor connections are indicated as 'T' in Fig. 7.4. The real-time SoC evaluation system performs the voltage, current and temperature measurements with the aid of a 16-bit Analog-to-Digital Converter (ADC).

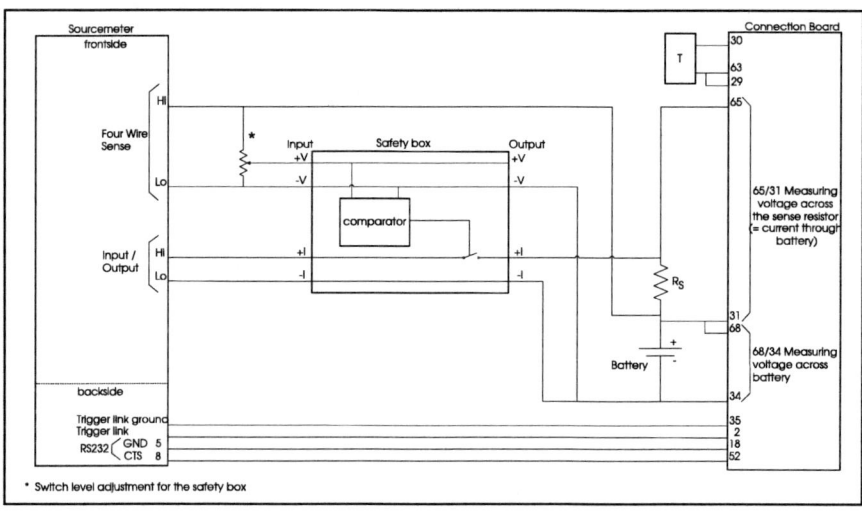

Fig. 7.4. Wiring diagram of the real-time SoC evaluation system.

The measurement accuracy of the SoC evaluation system

The maximum error in the voltage, current and temperature measurements will be further discussed in this section. The maximum error in the voltage measurement, ΔV_{max}, can be obtained by means of

$$|\Delta V_{max}| = |V_{off}| \pm \frac{1}{2}\frac{V_{fs}}{2^N} \qquad (7.3)$$

where V_{off} denotes the maximum offset value, V_{fs} the full scale range of the ADC voltage and N the number of bits.

The measurement offset depends on the chosen ADC voltage range. A range of +/– 5 V that respects the voltage of Li-ion batteries was chosen for the voltage measurements [1]. In this range the ADC has a maximum offset value (V_{off}) of +/– 0.8 mV [14] and a maximum resolution error of 0.076 mV for $N = 16$ and $V_{fs} = 10$ V. So the maximum voltage error is 0.88 mV.

Typically, the current is measured by measuring the voltage across a sense resistor (R_S) connected in series with the battery. This measured current is integrated over time and used to determine the battery's SoC. The higher current levels require substantially lower R_S values and higher power dissipation ratings. The low R_S resistance value results in a very small voltage drop across the shunt, which must be measured in order to determine the charge and discharge current in the real-time SoC evaluation system. Since one of the real-time SoC evaluation system's functions is to provide a time integration of the battery current in order to determine the battery's SoC, even small errors in the current measurement can cause substantial errors in the SoC measurement to accumulate over time (a common error when the signals to be measured are really small is offset of the current measurement device).

A range of +/– 2.5 A was chosen for the current measurements. This range corresponds to the GSM current maximum range. A voltage range of +/– 50 mV was chosen for the ADC to measure this current range. R_S consequently equals 20 mΩ. In this situation the ADC has an offset value V_{off} of +/– 0.029 mV [14]. In the case of $N = 16$ and $V_{fs} = 100$ mV the maximum voltage error of current measurements calculated using Eq. (7.3) will consequently be 0.03 mV.

A National Semiconductor precision temperature sensor with an accuracy within $\pm 1°C$ was used for the temperature measurements (see Fig. 7.4). In conclusion, the described real-time SoC evaluation system is capable of providing all the functions, needed to test the SoC algorithm, i.e. the voltage, current and temperature measurements and the application of the charge/discharge currents.

7.3 Results obtained with the algorithm using fresh batteries

A first set of tests was carried out with the described real-time SoC evaluation system to test the SoC algorithm accuracy and the SoC evaluation system validity. The following parts of the SoC algorithm were to this end implemented in the real-time SoC evaluation system: EMF_m as a function of temperature, V_{pm}, η_m and Q_{max} adaptation.

The tests were carried out using full and partial charge/discharge cycles at different constant C-rate currents and temperatures. During the full charge/discharge cycles the battery was each time fully charged to 4.2 V with the normal Constant-Current-Constant-Voltage (CCCV) charging method at 0.5 C-rate current in the Constant-Current (CC) mode. In the CV mode the voltage was kept constant at 4.2 V until the current reached a 0.05 C-rate value. Each charge period was followed by a rest period of 2 hours. After this rest period a discharge step at

0.1 C-rate current was applied until the battery voltage reached 3 V. This procedure was repeated at different constant discharge currents of 0.25, 0.5, 0.75 and 1 C-rate.

During the partial charge/discharge cycles the battery was charged to different SoC levels with the normal Constant-Current charging method at 0.5 C-rate current. Each charge period was followed by a rest period of 2 hours. After this rest period a discharge step at 0.1 C-rate current was applied until the battery voltage reached 3 V. This procedure was repeated at different constant discharge currents of 0.25, 0.5, 0.75 and 1 C-rate. All the experiments were carried out using the same battery at 5, 25 and 45°C.

A Sony US18500G3 Li-ion battery was used in all the tests. At the time of testing the battery was fairly new, having undergone approximately 5 discharge/charge cycles. The battery capacity was learned in the first charge cycle by using the method described in section 7.2. This showed that the Li-ion battery used in the measurements discussed in this section had a maximum capacity of 1177 mAh. Table 7.1 summarises the experimental results.

The discharge C-rate current and temperature T in [°C] at which the tests were carried out are given in columns one and two, respectively. The initial SoC at the start of the experiments is referred to as SoC_{st}, and the SoC remaining in the battery when the battery voltage equals V_{EoD} calculated by means of the overpotential and EMF model (see section 7.2) is expressed as SoC_l. These values are given in columns three and four, respectively. Columns five and six denote the remaining run-time predicted at the start of the experiment, t_{rstp} and the remaining run-time measured during the full discharge state, t_{rstm}, in minutes. The error in the remaining run-time, t_{re}, at the end of the experiment, the relative error in the remaining run-time, t_{rre}, and the SoC at the end of the experiment, SoC_{end}, are given in columns seven, eight and nine. SoC_{end} was calculated on the basis of SoC_{st}, Q_{max} and the capacity retrieved from the battery during the full discharge state. More information on SoC calculation during discharging will be given in section 7.4.2. The remaining run-time error, t_{re}, equals the remaining run-time value calculated by the real-time SoC evaluation system at the 3 V End-of-Discharge voltage level. The relative error in the remaining run-time is defined as follows

$$t_{rre}[\%] = 100 \frac{t_{re}}{t_{rstp} - t_{re}} \qquad (7.4)\ [15]$$

Two examples from Table 7.1 will be further explained below. The first example concerns the partial discharge at 0.25 C-rate current, 45°C and 58.4% SoC_{st}. In this situation the SoC indicator makes a pessimistic estimation. At the beginning of the discharge step the SoC evaluation system indicated 142.8 minutes remaining run-time. After 146.4 minutes the battery reached the level of 3 V. This means that the inaccuracy of the SoC system was −3.6 minutes in remaining run-time. In this example the relative error in remaining run-time t_{rre} is −2.5%.

Table 7.1. Results obtained with the real-time SoC evaluation system using a fresh US18500G3 Li-ion battery at different C-rate currents, temperatures and initial SoCs.

C-rate current	T [°C]	SoC_{st} [%]	SoC_1 [%]	t_{rstp} [min]	t_{rstm} [min]	t_{re} [min]	t_{rre} [%]	SoC_{end} [%]
0.10	5	97.1	2.1	609.9	616.9	−7.0	−1.1	1.0
		14.0	2.4	74.5	54.2	20.3	37.5	5.3
		4.1	2.1	12.8	12.5	0.3	2.4	2.2
	25	96.5	0.3	617.6	629.5	−11.9	−1.9	0.3
		40.3	2.1	245.2	199.2	46.0	23.1	9.4
		20.4	2.2	116.8	95.9	20.9	21.8	5.4
	45	99.0	2.3	620.8	642.4	−21.6	−3.4	−1.0
		58.0	1.8	360.8	371.2	−10.4	−2.8	0.7
		42.2	2.3	256.2	254.8	1.4	0.5	2.5
0.25	5	97.2	2.5	243.2	242.4	0.8	0.3	2.9
		5.6	2.5	8.0	4.6	3.4	73.9	3.8
		4.0	2.5	3.9	1.8	2.1	116.7	3.3
	25	98.4	2.6	246	251.5	−5.5	−2.2	0.4
		40.3	2.6	96.8	77.0	19.8	25.7	10.4
		20.8	2.7	46.5	37.5	9.0	24.0	6.2
	45	99.0	2.8	247.0	254	−7.0	−2.8	0.0
		58.4	2.8	142.8	146.4	−3.6	−2.5	1.4
		42.0	2.7	100.9	99.6	1.3	1.3	3.3
0.50	5	96.7	3.2	120.0	119.1	0.9	0.8	3.9
		5.6	3.0	3.3	0.5	2.8	560.0	5.2
		4.0	3.0	1.3	0.1	1.2	1200.0	3.9
	25	98.3	3.2	122.1	124.7	−2.6	−2.1	1.2
		40.2	3.2	47.5	37.3	10.2	27.3	11.2
		20.9	3.2	22.7	17.9	4.8	26.8	7.0
	45	99.0	3.3	122.9	126.7	−3.8	−3.0	0.2
		58.2*	3.4	70.4	72.2	−1.8	−2.5	2.0
		42.0	3.4	49.6	48.9	0.7	1.4	3.9
0.75	5	97.5	3.7	80.3	78.9	1.4	1.8	5.4
		5.6	3.6	1.7	0.1	1.6	1600	5.5
		4.2	3.5	0.6	0.0	0.6	100.0	4.1
	25	98.3*	3.9	80.8	82.4	−1.6	−1.9	1.9
		40.2	3.9	31.1	24.1	7.0	29	12.0
		20.7	3.6	14.6	11.3	3.3	29.2	7.7
	45	99.0	3.9	81.4	84.1	−2.7	−3.2	0.7
		58.2	4.0	46.4	47.7	−1.3	−2.7	2.4
		41.8	3.9	32.4	32.0	0.4	1.3	4.4
1.00	5	97.6	4.3	59.9	58.8	1.1	1.9	6.1
		5.4	4.0	0.9	0.0	0.9	100.0	5.4
		3.5	3.5	0.0	0.0	0.0	0.0	3.2
	25	98.4	4.5	60.3	61.6	−1.3	−2.1	2.4
		40.2	4.4	23.0	17.8	5.2	29.2	12.6
		23.4	4.2	12.3	8.9	3.4	38.2	9.6
	45	99.0	4.5	60.7	62.8	−2.1	−3.3	1.2
		58.0	4.6	34.3	35.3	−1.0	−2.8	2.9
		41.6	4.5	23.8	23.4	0.4	1.7	5.1

*Test examples considered for the SoC and remaining run-time uncertainty calculation discussed in sections 7.4.2 and 7.4.3.

In the second example an SoC_{st} value of 14.0% was indicated at the beginning of a full discharge cycle performed at 0.1 C-rate current and 5°C. In this situation the SoC indicator made a too optimistic estimation. At the beginning of the discharge step the system indicated 74.5 minutes remaining run-time. After 54.2 minutes the battery reached the level of 3 V. The inaccuracy of the SoC system was consequently 20.3 minutes in remaining run-time, whereas the relative error in the remaining run-time was 37.5%.

An important observation relates to the error in remaining run-time as a function of SoC_{st}. It follows from Table 7.1 that the error in remaining run-time increased when the experiment began at lower SoC_{st} values. This can be attributed mainly to the EMF hysteresis effect, which was not included in the SoC algorithm considered for these tests.

It follows from Table 7.1 that the calculated remaining run-time relative error sometimes exceeds 100%. For instance, a 1200% relative error was obtained for the partial discharge performed at 0.5 C-rate current and 5°C. At the beginning of the discharge step the system indicated 1.3 minutes remaining run-time. After 0.1 minute the battery reached the level of 3 V. The reason for this high calculated error is that the absolute error is much greater than the actual error. In this case the error in remaining run-time amounted to 1.2 minutes, which may be considered acceptable from a practical point of view. For this reason a new goal was defined for this work: *prediction of the remaining run-time (t_r) of any Li-ion battery with an uncertainty of less than 1 minute when t_{rstm} has a value lower than 100 minutes and an indication with an error of 1% or less when t_{rstm} has a value higher than 100 minutes under all realistic user conditions, including a wide variety of load currents and a wide temperature range.* This goal proved a difficult challenge.

The error in remaining run-time expressed in minutes by t_{re}, as a function of t_{rstp} [min.] is given in Fig. 7.5 to show just how close the measured data presented in Table 7.1 are to the actual data. t_{rstp} has been converted to a logarithmical scale and only error values lower than 25% (see Table 1) have been included. The accuracy limit values for the goal specified above are indicated by dashed lines in Fig. 7.5. For example, to satisfy the newly defined goal the modulus of t_{re} must be less than or equal to 7 minutes at t_{rstp} = 700 minutes.

It can be concluded from Table 7.1 and Fig. 7.5 that an always better than 20.3 minutes and 23.1% accuracy in the remaining run-time is obtained. These results do not meet the goal of this work. For a more accurate SoC and remaining run-time calculation the sources of the errors in the SoC evaluation system must be identified. The general remarks concerning uncertainty presented in section 3.5 will therefore be applied to the real-time SoC evaluation system described above.

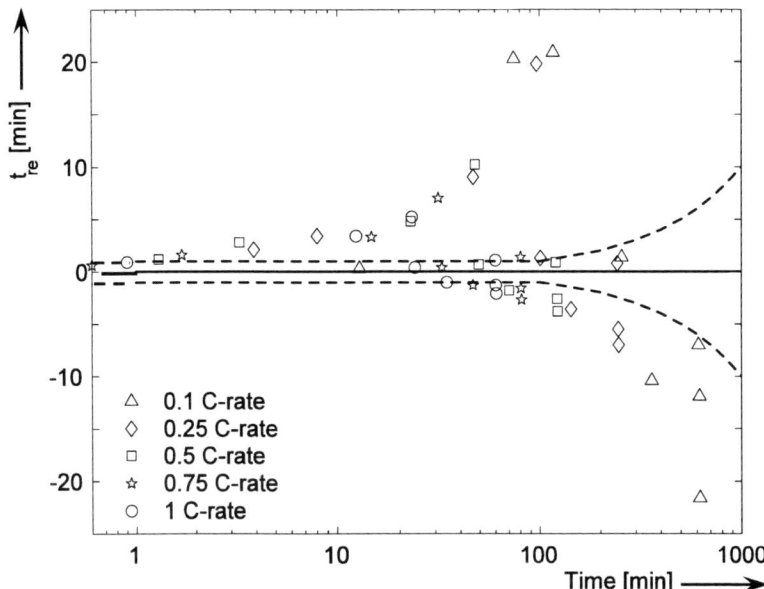

Fig. 7.5. The relative error in remaining run-time as a function of time converted to a logarithmical scale.

7.4 Uncertainty analysis

No matter what precautions are taken, there will always be a difference between a quantity's actual (but unknown) value and the result obtained in measurement. For an accurate SoC and t_r calculation the sources of error in each step in the SoC evaluation system must be identified and minimised. In this book 'measurement uncertainty' is a qualification of the expected closeness of a measurement result to the actual value [15].

7.4.1 Uncertainty in the real-time SoC evaluation system

The sources of error in each step in the SoC evaluation system will now be identified.

Initial state

In the initial state a battery's SoC is calculated by means of V and T measurements and the stored EMF_m. Accurate EMF-SoC curves have been obtained with a Maccor battery tester [16]. They were implemented in the SoC evaluation system using the physical EMF-SoC model (see chapter 4). The voltage (V_M), current (I_M), temperature (T_M) and time (t_M) measured with the Maccor are regarded as reference measurements. It was observed in repeated measurements and tests that the errors introduced in the EMF measurements by the spread between batteries and the EMF fitting (EMF_f) and detection (EMF_{dt}) methods used during the Maccor measurements (see chapter 4) are actually very small. They were therefore ignored in the SoC and t_r uncertainties calculation. The tests presented in this chapter were carried out using only fresh batteries. Errors introduced by the aging effect in the EMF-SoC and overpotential models were therefore also ignored.

The results obtained by applying the EMF model to an Li-ion battery as a function of three temperatures were presented in section 4.4. It was shown that the modelled EMF curve shows a good fit with the measured EMF curve obtained with the reference battery tester at 5, 25 and 45°C. The SoC error (SoC_e) obtained for the discharge EMF had a maximum value of 1.2% SoC (see Fig. 4.20).

In the SoC evaluation system the battery voltage is assumed to have a stable EMF in the initial state. However, when the SoC evaluation system is switched on, a current larger in module than the limit current may flow into or out of the battery or may just have just flown into or out of it. This means that an unstable voltage measurement may introduce, an error in the SoC_i calculation. The SoC evaluation system presented in this chapter was designed with a bipolar 16-bit Analog-to-Digital Converter (ADC) for measuring voltage. The maximum voltage error is 0.88 mV (see Eq. (7.3)). The voltage measurement will therefore also introduce an error in the SoC_i calculation. In repeated measurements and tests it was observed that the errors introduced in the SoC evaluation system's temperature and time (t) measurements are actually very small. They were therefore further ignored in the SoC and t_r uncertainties calculation. From the situations described above it can be concluded that in the initial state errors will be introduced in the SoC_i calculation by the *voltage* measurements and the modelling inaccuracy of EMF_m.

Standby state

A battery is assumed to be in equilibrium in the standby state and the SoC, expressed as SoC_s, is calculated by means of V and T measurements and the stored EMF_m. However, during the standby state a 1mA standby current is drawn from the battery. This means that an error will be introduced in the SoC_s calculation due to unstable voltage measurement. As in the initial state, the V measurement and the modelling inaccuracy of EMF_m will also introduce an error in the SoC_s calculation.

Backlight-on state

In the backlight-on state the SoC is determined on the basis of V and T measurements and the stored EMF_m. However, during the backlight-on state a 6 mA backlight-on current is drawn from the battery. Again an error will be introduced in the SoC_b calculation due to unstable voltage measurement. So the voltage measurements and the inaccuracy in EMF_m will introduce errors in the SoC_b calculation.

Transitional state

In the transitional state the voltage is not stable and SoC_t is calculated by means of voltage and time measurements, V_{pm} and EMF_m. However, during the transitional state a small current, *i.e.* I_s, is drawn from the battery. This means that an error will again be introduced in the SoC_t calculation due to unstable voltage measurement. So errors will be introduced in the SoC_t calculation by the V measurements, V_{pm} and the modelling inaccuracy of EMF_m.

It can be concluded from the situations described above that the main errors in SoC calculation during the initial, standby, backlight-on and transitional states are the errors introduced in *voltage* measurement and those generated by the modelling inaccuracy of EMF_m and V_{pm}.

Charge state

In the charge state a positive current flows into the battery. The SoC in the charge state, SoC_{ch}, is determined by means of cc. The SoC evaluation system presented in this chapter was designed with a bipolar 16-bit Analog-to-Digital Converter (ADC) for measuring the current. The current is determined by measuring the voltage drop across a 20mΩ sense resistor connected in series with the battery (see Fig. 7.4). The maximum voltage error in current measurements equals 0.03 mV (see Eq. (7.3)). In this case the current error measured when the sense resistor value is calibrated will be 1.5 mA. To allow Q_{ch} [mAh] measured by means of cc to be translated into SoC_{ch} [%], Q_{max} must be known. The current measurement and integration and the Q_{max} measurements will consequently introduce errors in the SoC_{ch} calculation.

Discharge state

In the discharge state a battery is discharged and a negative current larger in module than I_{lim} flows out of the battery. SoC_d is determined by means of cc and Q_{max} calculations, which are also the main sources of errors in the SoC_d calculation. In addition to SoC_d, t_r available under the discharge conditions is also calculated during the discharge state. As shown by Eq. (7.2) the remaining run-time is inferred from SoC_d, SoC_l, Q_{max} and I_d. Therefore, in addition to simple cc, the effect of a battery's overpotential must also be considered in the calculation of t_r. Accurate overpotential curves obtained with a Maccor battery tester were implemented in the real-time SoC evaluation system using the overpotential model (η_m). The results obtained by applying the overpotential model to an Li-ion battery as a function of four C-rate currents were presented in section 5.2. As shown in that section, the modelled overpotential curve yields an SoC_l error of 0.4%. In repeated measurements and tests it was observed that the errors resulting from spread in the overpotential model calculation are actually very small; they were therefore ignored in the remaining run-time uncertainty calculation.

From the situations described above it can be concluded that the main error in SoC calculation during the charge and discharge states is the error generated in cc, whereas the main errors in the t_r calculation during the discharge state are the errors introduced by SoC_d, SoC_l, Q_{max} and the current measurement. In summary, the SoC and t_r calculation uncertainties depend on the state in which the SoC evaluation system operates.

Fig. 7.6 presents a diagram illustrating the sources of error in the SoC-evaluation system. The single-line blocks represent the errors introduced by the on-line measurements in the SoC evaluation system, the grey blocks the errors generated by the battery models and off-line measurements, while the dashed and double-line blocks represent errors that have not been considered or have been ignored, and errors resulting from t_r uncertainties calculation, respectively. Finally, the circles denote errors introduced by the variables measured on-line that were used as input for the battery model calculations.

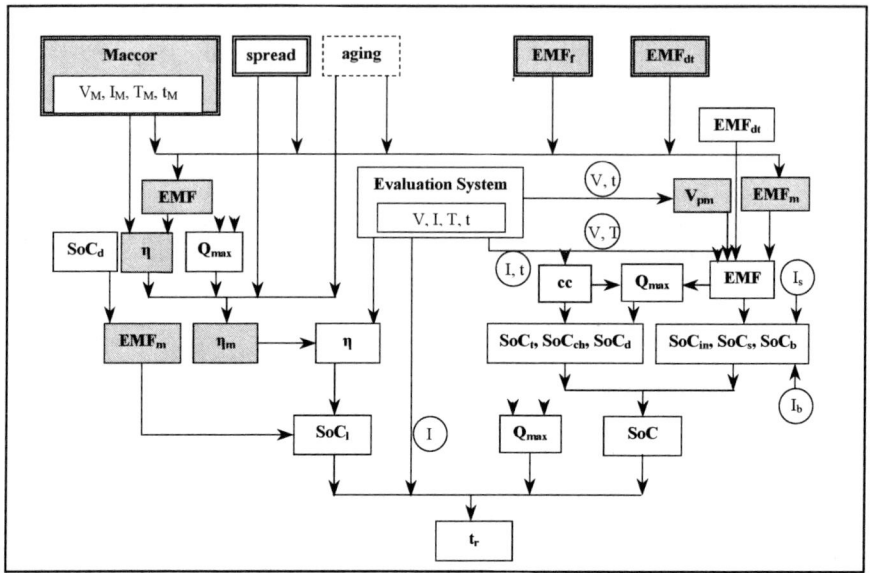

Fig. 7.6. Schematic representation of the sources of error in the real-time SoC evaluation system.

7.4.2 The SoC uncertainty

Two examples from section 7.3 at different C-rate currents, SoC_{st} and temperatures will be further considered for the SoC uncertainty calculation.

In the first example the SoC evaluation system runs through the following sequence of states: initial, standby, full charge, transitional, standby and full discharge. At the beginning of the test the SoC evaluation system was switched on and the initial state was entered (see Fig. 7.2). In this state a battery voltage of 3.01 V and a battery temperature of 25°C were measured. These measurements and the stored EMF_m yield an SoC_i value of 0%. After a short rest period in which the battery voltage and temperature were continuously monitored the SoC evaluation system was switched to standby state (see Fig. 7.2). In this state a SoC_s of 0% was calculated on the basis of V and T measurements and the stored EMF_m.

A 0.5 C-rate positive current was then applied to the battery and the SoC evaluation system switched to the charge state (see Fig. 7.2). During this state the battery was fully charged using the usual Constant-Current-Constant-Voltage (CCCV) charging method at a 0.5 C-rate charging current. In the CV mode the voltage was kept constant at 4.2 V until the current reached 0.05 C-rate. The SoC during the charge state (SoC_{ch}) was calculated as follows

$$SoC_{ch} = SoC_s + 100\frac{Q_{ch}}{Q_{max}} \qquad (7.5)$$

During the CCCV charging a Q_{ch} of 1175 mAh was determined on the basis of cc and an assumed maximum capacity Q_{max} of 1188 mAh based on previous maximum capacity measurements obtained for this type of battery. The SoC_{ch}

consequently amounted to 98.9% (see Eq. (7.5)). However, by the end of the CV mode at 25°C the SoC level should by definition be 100%. This meant that SoC_{ch} had to be calibrated and a new SoC_{ch} value of 100% had to be programmed. In this example the error in SoC_{ch} was consequently 1.1%. This SoC error was calculated as the difference between the actual SoC_{ch}, i.e. 100%, and the SoC_{ch} calculated using Eq. (7.5), i.e. 98.9.

The sources of error in the SoC_{ch} calculation will be further discussed below. According to Eq. (7.5), SoC_{ch} is a function of three parameters: SoC_s, Q_{max} and Q_{ch}. Each of these parameters contributes to the uncertainty in SoC_{ch}. The partial derivative of SoC_{ch} in Eq. (7.5) (and considering SoC_s in unit value for simplicity) yields

$$\frac{dSoC_{ch}}{SoC_{ch}} = \frac{1}{Q_{max}SoC_s + Q_{ch}} \left(Q_{max}SoC_s \frac{dSoC_s}{SoC_s} + Q_{ch}\frac{dQ_{ch}}{Q_{ch}} - Q_{ch}\frac{dQ_{max}}{Q_{max}} \right) \quad (7.6)$$

where $d(.)/(.)$ denotes the relative error in the quantity (.). Each of these relative errors will now be briefly considered to estimate their influence on the overall uncertainty.

After investigation and comparison with a reference current measurement method it was concluded that the Q_{ch} measurement introduces a 0.2% SoC relative error in SoC_{ch}.

The error introduced by SoC_s was found to be very small and was therefore ignored in the SoC_{ch} uncertainty calculation.

The third parameter, Q_{max}, satisfies Eq. (7.1). So the error in Q_{max} follows from the errors in SoC_s and Q_{ch}. The partial derivative of Q_{max} in Eq. (7.1) results in

$$\frac{dQ_{max}}{Q_{max}} = \frac{dQ_{ch}}{Q_{ch}} + \frac{1}{SoC_{sf} - SoC_{si}} \left(SoC_{si}\frac{dSoC_{si}}{SoC_{si}} - SoC_{sf}\frac{dSoC_{sf}}{SoC_{sf}} \right) \quad (7.7)$$

where $d(.)/(.)$ denotes the relative error in the quantity (.).

It can be concluded from Eq. (7.7) and the situation described above that the main error in maximum capacity is introduced by the assumed maximum capacity value, i.e. 1188 mAh. Therefore Q_{max} was recalculated and a new value of 1177 mAh was programmed in the SoC evaluation system. It should be noted that other initial values could also be chosen for the maximum capacity. The same error calculation principle will then of course apply.

The charging was followed by a rest period of about 4 hours, during which a small current I_s of 1 mA was drawn from the battery. At the beginning of the rest period the SoC evaluation system switched to the transitional state (see Fig. 7.2). To allow comparison, the SoC value during the transitional state was calculated using cc and also on the basis of the V, T, t, V_{pm} and EMF_m measurements and calculations. The SoC during the transitional state, SoC_t, was calculated by means of cc as follows

$$SoC_t = SoC_{ch} - 100\frac{Q_d}{Q_{max}} \quad (7.8)$$

where Q_d was determined by means of cc.

It follows from Eq. (7.8) that SoC_t is a function of three parameters: SoC_{ch}, Q_{max} and Q_d. Each of these parameters contributes to the uncertainty in SoC_t. It should be noted that the values of the first two parameters, *i.e.* SoC_{ch} and Q_{max}, were calibrated at the end of the charge step. These two sources of error were therefore ignored in the SoC_t uncertainty calculation. During the first two hours of the rest period an accurate charge Q_d of about 2 mAh was measured by means of cc, and an SoC_t value of about 99.8% was obtained at the end of the transitional state (see Eq. (7.8)). It can be concluded that the errors in the SoC_t calculation introduced by the third parameter, *i.e.* Q_d, may also be ignored.

During the transitional state SoC_t was also calculated on the basis of the V, T, t, V_{pm} and EMF_m measurements and calculations. The V and t measurements and V_{pm} yielded a voltage of 4.177V. This predicted voltage was used as input for the stored EMF_m. This resulted in an SoC_t of 99%.

After a two-hour rest period the SoC evaluation system switched to the standby state. At the beginning of the standby state a battery voltage of 4.177 V and a battery temperature of 25°C were measured. These measurements and the stored EMF_m yield an SoC_s of 99%. From the situations described above it can be concluded that the SoC_s calculation after the two-hour rest period had an error of 0.8% SoC. Another conclusion relates to V_{pm}. The EMF value predicted during the transitional state, *i.e.* 4.177 V, equals the EMF value at the beginning of the standby state. So the voltage-prediction model does not introduce any errors in the SoC_t calculation.

At the end of the standby state a SoC_s of 98.3% was calculated on the basis of the V, T, t and EMF_m measurements and calculations, whereas the actual SoC_s was 99.7%. The actual SoC_s value was calculated by means of cc. It can be concluded that the error introduced by EMF_m contributed most to the SoC_s error and amounted to –1.4% SoC.

A 0.5 C-rate negative current was then drawn from the battery and the SoC evaluation system switched to the discharge state (see Fig. 7.2). During this state the battery was discharged until the voltage reached a defined End-of-Discharge Voltage (V_{EoD}) of 3 V at 0.5 C-rate discharging current. The SoC during the discharge state, SoC_d, was calculated using

$$SoC_d = SoC_s - 100 \frac{Q_d}{Q_{max}} \qquad (7.9)$$

Eq. (7.9) shows that SoC_d is a function of SoC_s, Q_{max} and Q_d. So SoC_s and Q_d will introduce errors in SoC_d, because Q_{max} has been calibrated. During the discharge state a Q_d of 1143 mAh was determined by means of cc. At the end of the discharge step an SoC_d of 1.2% was calculated (see Table 7.1), whereas the actual SoC_d was 2.6%. The actual SoC_d was calculated by assuming the actual SoC_s of 99.7%.

The sources of error in the SoC_d calculation will now be further discussed. By assuming the difference between SoC_d and SoC_s to be ΔSoC_d and assuming absolute values for simplicity, Eq. (7.9) becomes

$$\Delta SoC_d = 100 \frac{Q_d}{Q_{max}} \tag{7.10}$$

It follows from Eq. (7.10) that the absolute error in SoC_d will be the sum of the absolute errors in SoC_s and ΔSoC_d. The measurements showed that the error introduced by Q_d during discharging was actually very small, so it was further ignored in the SoC_d uncertainty calculation. It should also be noted that the Q_{max} value was calibrated. This means that the relative error in SoC_d was introduced mainly by SoC_s and equalled -1.4%.

7.4.3 The remaining run-time uncertainty

The SoC evaluation system presented in this chapter also calculates the remaining run-time available under the discharge conditions (see Eq. (7.2)). The remaining run-time calculation method is schematically illustrated in Fig. 7.7.

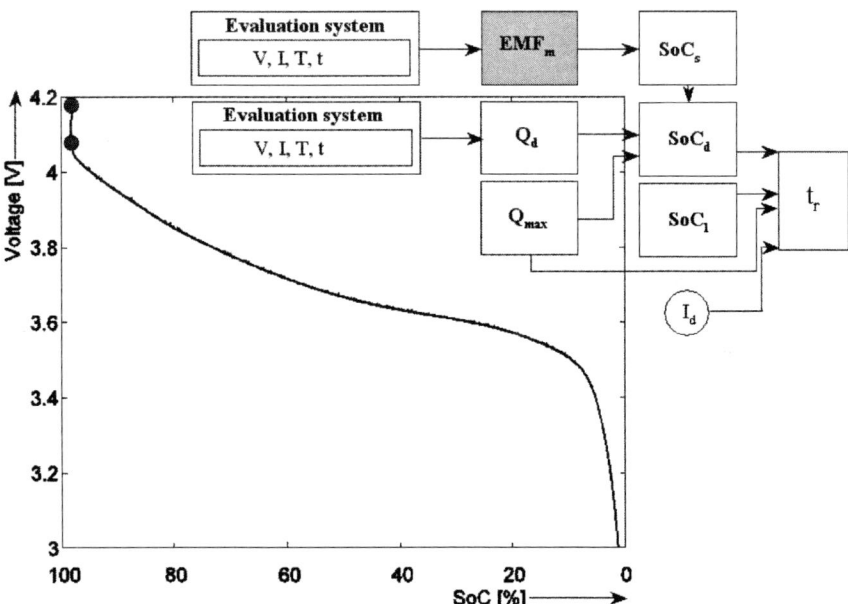

Fig. 7.7. Schematic representation of the calculation of the remaining run-time available under the discharge conditions (see also Fig. 7.6).

To allow t_r to be calculated, an SoC_l value is calculated at the beginning of the discharge state on the basis of the overpotential prediction and EMF_m. In this example a SoC_l of 3.2% was calculated, whereas the actual SoC_l amounted to 2.6%. The actual SoC_l was calculated by considering the actual SoC_d at the end of discharging (see section 7.4.2). As a result, t_r inferred from Eq. (7.2) using the measured SoC_d (98.3%), SoC_l (3.2%), Q_{max} (1177 mAh) and I_d at the beginning of discharging (see Fig. 7.7), i.e. 0.55 A, was 122.1 minutes, whereas the actual t_r was 124.7 minutes. It should be added that the SoC_d value at the beginning of discharging corresponded to the SoC value of the previous state.

To obtain a high level of accuracy in the remaining run-time interpretation the actual t_r was calculated in two different ways. First of all the remaining run-time value measured by the real-time SoC evaluation system at the defined 3 V End-of-Discharge voltage level was assumed to be the true value. Then the actual SoC_d and SoC_l values were used to calculate the actual remaining run-time value. So in this example the t_r calculation had an error of –2.6 minutes.

We will now consider the sources of error in the t_r calculation. Taking the partial derivative of t_r and considering t_r in hours, SoC dimensionless, Q_{max} in [mAh] and I_d in [mA], Eq. (7.2) results in

$$t_{rre} = t_r \left(\frac{dQ_{max}}{Q_{max}} - \frac{dI_d}{I_d} \right) + \frac{Q_{max}}{I_d} \left(dSoC_d - dSoC_l \right) \tag{7.11}$$

where t_{rre} denotes the relative error in t_r, dI_d/I_d the relative error in I_d and $dSoC_d$ and $dSoC_l$ the error in SoC_d, i.e. 0.014, and in SoC_l, i.e. –0.007, respectively. It should be noted that $dSoC_l$ was calculated as a difference between the true and the measured SoC_l values. For this calculation the second number after the decimal point has been also taken into consideration.

Eq. (7.11) shows that Q_{max}, I_d, SoC_d and SoC_l contribute to the uncertainty in t_r. The measurements showed that the error introduced by I_d during discharging is very small. The errors introduced by Q_{max} and I_d were therefore ignored in the t_r uncertainty calculation.

It follows from Eq. (7.11) that the error in t_r is –0.045 hour or –2.7 minutes. This calculated error corresponds well to the error obtained in the example. It can be concluded from the situations described above that the main errors in t_r result from errors in SoC_d, i.e. 1.8 minutes, and in SoC_l, i.e. –0.9 minute. These two error values have opposite signs, which mean that they do not (partly) cancel.

In the second example the SoC evaluation system runs through initial, standby, partial charge, transitional, standby and partial discharge states. At the beginning of the test the SoC evaluation system was switched on and the initial state was entered (see Fig. 7.2). In this state a battery voltage of 3.03 V and a battery temperature of 45°C were measured. These measurements and the stored EMF_m yielded an SoC_i value of 0.5%.

After a short rest period in which the battery voltage and temperature were continuously monitored the SoC evaluation system switched to the standby state (see Fig. 7.2). In this state a SoC_s of 0.5% was calculated on the basis of V and T measurements and the stored EMF_m.

A 0.5 C-rate positive current was then applied to the battery and the SoC evaluation system switched to the charge state (see Fig. 7.2). During this state the battery was partially charged to 3.98 V using the usual Constant-Current (CC) charging method at a 0.5 C-rate. The SoC during the charge state (SoC_{ch}) was calculated on the basis of SoC_s, Q_{ch} and Q_{max} (see Eq. (7.5)). During the CC charging a Q_{ch} of 675 mAh was measured by means of cc, resulting in an SoC_{ch} of 57.8% (see Eq. (7.5)). After investigation and comparison with a reference measurement it was concluded that the Q_{ch} measurement introduced an 0.2% SoC relative error in the SoC_{ch} calculation. The error introduced by SoC_s was found to be very small and was further ignored in the SoC_{ch} uncertainty calculation. In this example the error in SoC_{ch} was 0.2%. This SoC error was calculated as the difference between the actual SoC_{ch}, i.e. 58.0%, and the SoC_{ch} calculated using Eq. (7.5), i.e. 57.8%.

The charge step was followed by a rest period of about 2 hours during which a small current I_s of 1 mA was drawn from the battery. At the beginning of the rest period the SoC evaluation system switched to the transitional state (see Fig. 7.2). To allow comparison, the SoC during the transitional state was calculated by means of cc and also on the basis of the V, T, t, V_{pm} and EMF_m measurements and calculations. During the two-hour rest period a charge Q_d of about 2 mAh was measured by means of cc, and a value of about 57.6% was obtained for SoC_t at the end of the transitional state (see Eq. (7.8)). It can be concluded that the error introduced in the SoC_t calculation by cc may be ignored.

During the transitional state SoC_t was also calculated on the basis of the V, T, t, V_{pm} and EMF_m measurements and calculations. The V and t measurements and the voltage-prediction model yielded a value of 3.854V. This predicted voltage was used as input for the stored EMF_m. The SoC system consequently calculated a SoC_t value of 58.4%.

After one hour's rest the SoC evaluation system switched to the standby state. At the beginning of the standby state a battery voltage of 3.854 V and a battery temperature of 45°C were measured. These measured values lead to an SoC of 58.4%. It follows from this example that the error in SoC_s amounted to 0.8%. The error in SoC_s was calculated as the difference between the actual SoC_s calculated by means of cc and the SoC_s calculated on the basis of the V, T, t, V_{pm} and EMF_m measurements and calculations. This SoC error was attributed to the influence of hysteresis in the EMF_m on the basis of the good accuracy obtained in EMF_m simulations at 58.4% SoC and 45°C (see Fig. 4.20). Another conclusion relates to the V_{pm}. From the situations described above it follows that V_{pm} does not introduce any errors in the SoC_t calculation. So the error introduced by the EMF_m, in particular the fact that hysteresis was not taken into account, contributed most to the SoC_s error.

An 0.25 C-rate negative current was then drawn from the battery and the SoC evaluation system switched to the discharge state (see Fig. 7.2). During this state the battery was discharged to the End-of-Discharge Voltage level. During the discharge state a Q_d of 671 mAh was determined by means of cc. At the end of the discharge step an SoC_d value of 1.4% was calculated, whereas the actual SoC_d amounted to 0.6%. The actual SoC_d was calculated by assuming the actual SoC_s to be 57.6%.

We will now consider the sources of error in the SoC_d calculation. It follows from Eqs. (7.9) and (7.10) that the absolute error in SoC_d will be the sum of the absolute errors in SoC_s and ΔSoC_d. The measurements showed that the error introduced by Q_d during discharging was very small; it was therefore further ignored in the SoC_d uncertainty calculation. The relative error in the SoC_d was therefore introduced mainly by SoC_s and equalled 0.8%.

To allow t_r to be calculated, an SoC_l value of 2.8% was calculated at the beginning of the discharge state, whereas the actual SoC_l amounted to 0.6%. The actual SoC_l was calculated by considering the actual SoC_d at the end of discharging. t_r inferred from the measured SoC_d (58.4%), SoC_l (2.8%), Q_{max} (1177 mAh) and I_d at the beginning of discharging, i.e. 0.275 A, was consequently 142.8 minutes, whereas the actual t_r was 146.4 minutes. In this example the t_r calculation had an error of −3.6 minutes.

Let us now consider the sources of error in the t_r calculation. The measurements showed that the error introduced by I_d during discharging was very small; it was therefore ignored in the t_r uncertainty calculation. From the situations described above and Eq. (7.11) it follows that the error in t_r is -0.06 hour or -3.6 minutes. This calculated error corresponds well to the error obtained in the example. It can be concluded that the main errors in t_r resulted from errors in SoC_d, i.e. 2 minutes, and in SoC_l, i.e. 5.6 minutes. The errors introduced by SoC_d and SoC_s in the t_r have the same sign, and must therefore be subtracted from one other (see Eq. (7.11)).

From the examples considered in this section it can be concluded that accurate modelling of the EMF=f(SoC) relationship (see chapter 4) and of the overpotential (see chapter 5) is not enough to obtain a high level of accuracy in remaining run-time prediction either.

7.5 Improvements in the new SoC algorithm

In this section new methods for modelling the Electro-Motive Force and predicting a battery's State-of-Charge-left (SoC_l) will be described. Results obtained with these methods will also be compared with results of the EMF=f(SoC) and overpotential methods described in chapters 4 and 5. The models will be developed on the basis of the conclusions of the error analysis performed in section 7.4.

7.5.1 A new State-of-Charge-Electro-Motive Force relationship

Accurate SoC fitting results were obtained with the EMF physical model developed in chapter 4. Although this model each time yielded a high level of accuracy – i.e. an error of less than 1.2% in SoC – for the discharge EMF (see Fig. 4.20), this does not seem to be sufficient to guarantee a high level of accuracy in remaining run-time indication, too (see Table 7.1). It should be noted that for an accurate description of the EMF-SoC relationship, the hysteresis effect on the EMF, discussed in chapter 4, should also be considered. This section therefore presents a new method for determining SoC=f(EMF) in which the SoC during equilibrium is determined on the basis of the measured EMF and temperature without any need for mathematical inversion, which could compromise the SoC indication accuracy. The mathematical function contains a set of parameters that are found by fitting to available measured EMF curves. The advantage is that no numerical inversion is needed, as in the method discussed in [1]. The function describing the SoC=f(EMF) relationship is given by Eqs. (7.12), (7.13) and (7.14).

$$f_x = a_{10} + x + a_{11}|x|^{p_{11}} s_x^{q_{11}} + a_{12}|x|^{p_{12}} s_x^{q_{12}} \quad (7.12)$$

in which the dimensionless $x = F(E_o^x - \text{EMF})/\text{RT}$, F denotes the Faraday constant (96485 Cmol^{-1}), EMF[V] is the measured Electro-Motive Force, R the gas constant (8.314 J (mol K)$^{-1}$) and T the (ambient) temperature in [K]. $|x|$ denotes the absolute value of x and s_x denotes the sign of x. E_o^x in [V] and the dimensionless a_{10}, a_{11}, p_{11}, q_{11}, a_{12}, p_{12} and q_{12} are parameters obtained by fitting the measured battery EMF.

$$f_z = a_{20} + z + a_{21}|z|^{p_{21}} s_z^{q_{21}} + a_{22}|z|^{p_{22}} s_z^{q_{22}} \qquad (7.13)$$

in which the dimensionless $z = F(E_o^z - \text{EMF})/RT$, $|z|$ denotes the absolute value of z and s_z denotes the sign of z. E_o^z [V] and the dimensionless a_{20}, a_{21}, p_{21}, q_{21}, a_{22}, p_{22} and q_{22} are parameters obtained by fitting the measured battery EMF.

Using Eqs. (7.12) and (7.13), the SoC can be inferred from the measured EMF as follows

$$SoC = A\left[\frac{1-w}{1+e^{f_x}} + \frac{w}{1+e^{f_z}}\right] \qquad (7.14)$$

in which SoC denotes the dimensionless State-of-Charge and the dimensionless A and w are parameter values determined by fitting the measured battery EMF.

In order to include the influence of temperature in the SoC=f(EMF) relationship, linear dependence of each of the model parameters was assumed according to

$$par(T) = par(T_{ref}) + (T - T_{ref})\Delta par \qquad (7.15)$$

where T_{ref} is a reference temperature (e.g. 25°C), T is the ambient temperature and $par(T_{ref})$ is the value of one of the SoC=f(EMF) model parameters at temperature T_{ref}. The Δpar value is the sensitivity to temperature determined for each parameter $par(T_{ref})$.

7.5.2 A new State-of-Charge-left model

A new method for determining the remaining run-time will be proposed in this section. With this method it is not necessary to predict a battery's overpotential, measure a battery's voltage or use the EMF model under current flowing conditions. To this end a new predictive State-of-Charge-left (SoC_l) mathematical function was developed as follows

$$SoC_l = \frac{\left[C\left(\frac{SoC_{st}}{100}\right)^{\varsigma(\vartheta - C)}\right]^{\gamma + \delta T}}{\alpha + \beta T} \qquad (7.16)$$

where SoC_{st} [%] denotes the SoC at the beginning of discharge at C-rate current C and temperature T [°C]. β [T^{-1}], δ [T^{-1}] and the dimensionless α, γ, ς and ϑ are parameters fitted to measured SoC_l data.

When the SoC_l function described by Eq. (7.16) is substituted in Eq. (7.2) the remaining run-time at point B can be predicted immediately after discharging has started at point A (see Fig. 3.4). The function described by Eq. (7.16) contains a set of parameters that are found by fitting to available SoC_l measured values. The advantages are that SoC_l can be easily measured (see Fig. 3.4), that prediction of the overpotential, voltage measurement and EMF model calculation are not necessary under loading conditions and that the remaining run-time is calculated directly by only one function.

7.5.3 Determination of the parameters of the new models

The parameters used in Eqs. (7.12) – (7.15) were found by fitting the SoC=f(EMF) relationship to measured charge/discharge EMF curves obtained with a reference battery tester at three temperatures. The charge/discharge EMF curves were measured with the voltage-relaxation method described in chapter 6. The EMF model parameters are summarised in Table 7.2.

Column one gives the symbol of the EMF model parameter. The values of the EMF parameter obtained for the charge and discharge EMFs are given in columns two and three, respectively. Column five indicates the units of the EMF model parameters. The SoC=f(EMF) parameters presented in Table 7.2 were used in the simulations whose results are shown in Figs. 7.8 and 7.9.

Figs. 7.8 and 7.9 show that the modelled charge/discharge EMF curve used in the system shows a good fit with the measured discharge curve obtained with the reference battery tester at all temperatures. The maximum error in SoC appears to be 0.8%. In Fig. 4.20 this was 1.2 %, so the new method gives a better result.

The SoC_l parameters used in Eq. 7.17 will now be determined. To obtain information on SoC_l the battery was discharged from different SoC_{st} values at different constant C-rates and temperatures. The SoC value at the end of discharging was taken as the SoC_l value. Another way of determining the SoC_l value would be to apply the voltage-relaxation model and the newly developed SoC=f(EMF) relationship. This way the battery equilibrium voltage predicted after the first few minutes of the voltage-relaxation curve can be converted into an SoC_l value by using the SoC=f(EMF) function. Verification showed that the aforementioned methods lead to the same predicted SoC_l values. The measured SoC_l values were used as input in the SoC_l model (SoC_{lm}) described by Eq. (7.16). The SoC_{lm} model parameter values are presented in Table 7.3. Column one gives the symbol of the SoC_l model parameter. The values and units of the SoC_l model parameters are given in columns two and three.

Table 7.2. The battery SoC=f(EMF) model parameter values.

Parameter	Charge value	Discharge value	Unit
E_0	3.72	3.71	[V]
ΔE_0	0.00	$-3.52\ 10^{-4}$	$[V\ T^{-1}]^+$
a_{10}	2.42	2.04	[1]
Δa_{10}	$-4.98\ 10^{-2}$	$-4.26\ 10^{-2}$	$[T^{-1}]^+$
a_{11}	$2.28\ 10^{-1}$	$2.41\ 10^{-1}$	[1]
Δa_{11}	$2.28\ 10^{-3}$	$8.22\ 10^{-3}$	$[T^{-1}]^+$
a_{12}	$3.06\ 10^{-2}$	$2.05\ 10^{-2}$	[1]
Δa_{12}	$-3.83\ 10^{-4}$	$-2.46\ 10^{-4}$	$[T^{-1}]^+$
a_{20}	3.82	3.74	[1]
Δa_{20}	$2.11\ 10^{-3}$	$3.99\ 10^{-3}$	$[T^{-1}]^+$
a_{21}	$-7.59\ 10^{-1}$	$-7.59\ 10^{-1}$	[1]
a_{22}	$2.46\ 10^{-4}$	$6.77\ 10^{-4}$	[1]
p_{11}	1.29	1.35	[1]
Δp_{11}	$1.742\ 10^{-2}$	$3.58\ 10^{-3}$	$[T^{-1}]^+$
p_{12}	3.00	3.00	[1]
Δp_{12}	$1.86\ 10^{-2}$	$1.85\ 10^{-2}$	$[T^{-1}]^+$
p_{21}	1.06	1.06	[1]
Δp_{21}	$-8.49\ 10^{-5}$	$-8.01\ 10^{-5}$	$[T^{-1}]^+$
p_{22}	2.00	2.00	[1]
q_{11}	0.00	0.00	[1]
q_{12}	1.00	1.00	[1]
q_{21}	1.00	1.00	[1]
q_{22}	0.00	0.0	[1]
A	6.64	6.41	[1]
ΔA	$2.813\ 10^{-3}$	$3.27\ 10^{-3}$	$[T^{-1}]^+$
w_2	$9.56\ 10^{-1}$	$9.56\ 10^{-1}$	[1]
Δw_2	$-1.62\ 10^{-4}$	$-1.92\ 10^{-4}$	$[T^{-1}]^+$

[+] See Eq. (7.15).

Table 7.3. SoC-left model parameters.

Parameter	Value	Unit
ς	$5.90\ 10^{-6}$	–
ϑ	$7.25\ 10^{4}$	–
γ	$3.50\ 10^{-1}$	–
δ	$1.30\ 10^{-3}$	$[T^{-1}]^+$
α	$1.18\ 10^{-1}$	–
β	$3.75\ 10^{-3}$	$[T^{-1}]^+$

[+] See Eq. (7.15).

The SoC_l values obtained on the basis of SoC_{lm} and fitted SoC_l (SoC_{lf}) values (using Eq. (7.16)) are presented in Fig. 7.10. The difference between the measured and fitted SoC_l values is indicated in Fig. 7.11 to show how close the measured values are to the fitted data. It follows from Figs. 7.10 and 7.11 that the maximum difference between the measured and fitted SoC_l values equalled 0.6% at 45°C.

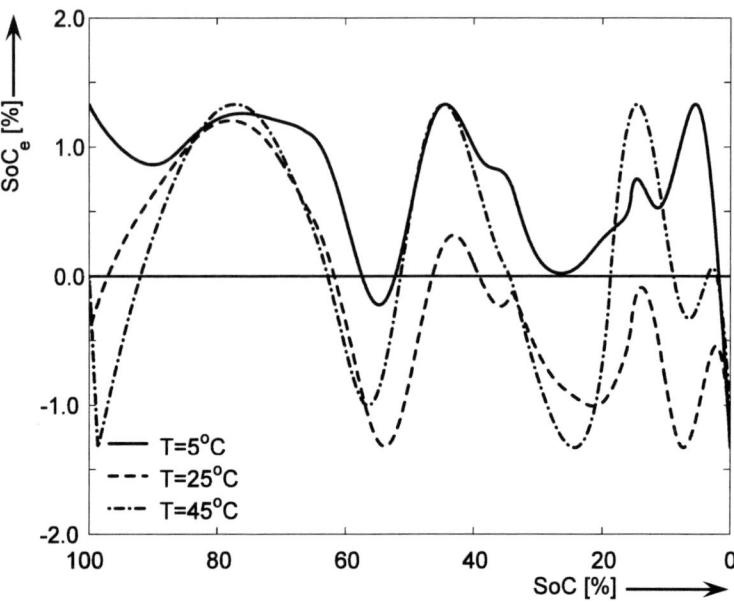

Fig. 7.8. Accuracy of SoC indication using the measured charge EMF curve versus the fitted EMF curve at 5, 25 and 45°C. The horizontal axis shows the SoC [%] normalised to maximum capacity.

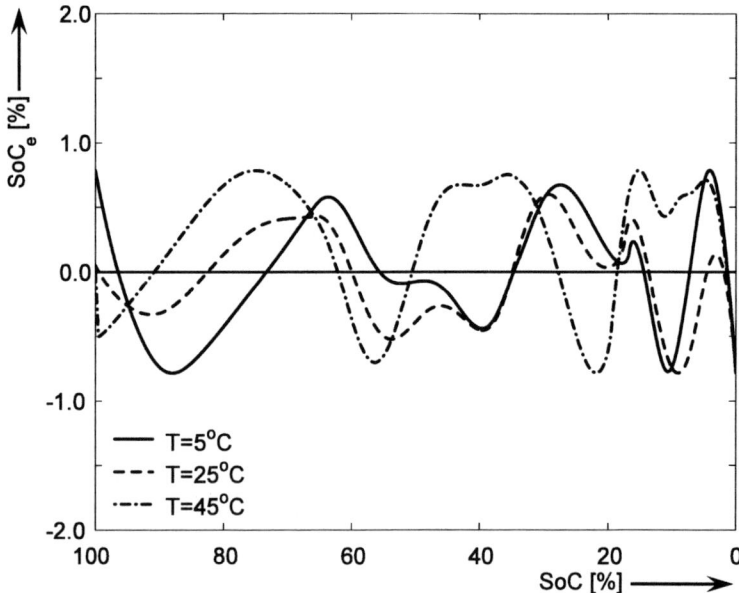

Fig. 7.9. Accuracy of SoC indication using the measured discharge EMF curve versus the fitted EMF curve at 5, 25 and 45°C as a function of the SoC [%] normalised to maximum capacity.

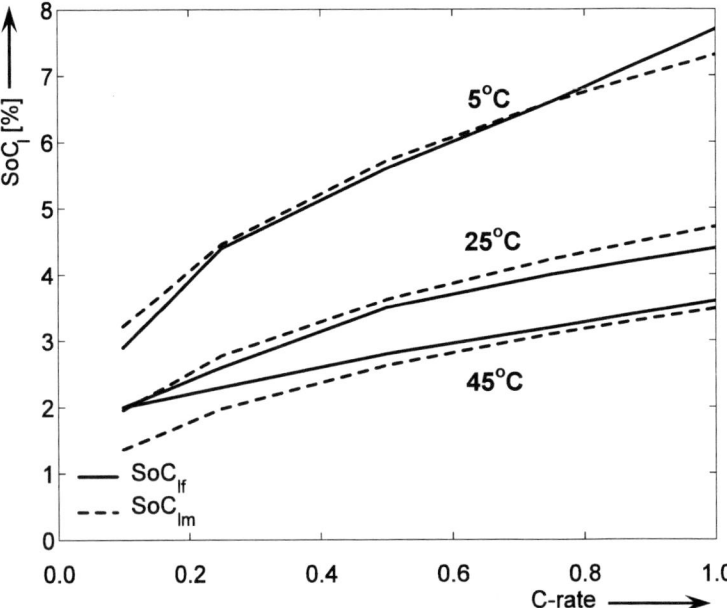

Fig. 7.10. Accuracy of SoC indication using the measured SoC_l values versus the fitted values at 5, 25 and 45°C. The horizontal axis shows the C-rate.

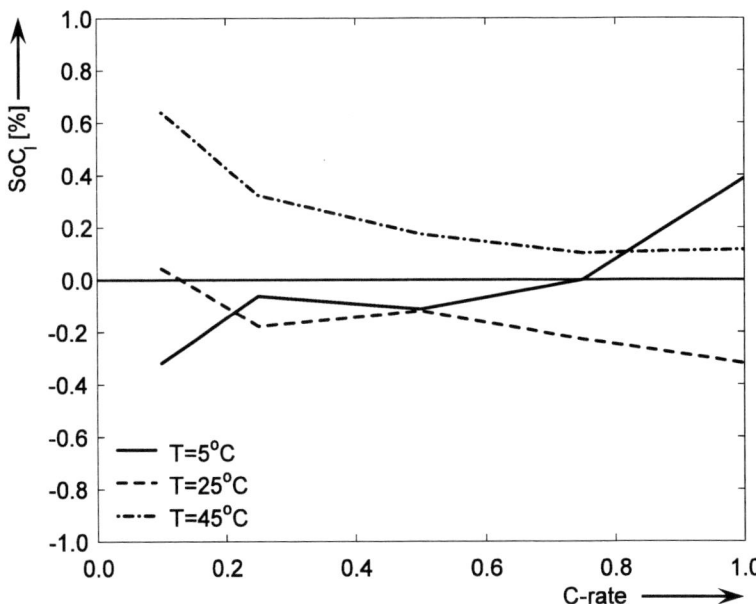

Fig. 7.11. The difference between the measured and fitted SoC_l values obtained during discharging at different currents at three temperatures (see also Fig. 7.10).

7.5.4 Test results

Two sets of tests at constant and switching C-rate currents were carried out with the improved real-time SoC evaluation system in order to demonstrate the SoC algorithm accuracy and the SoC evaluation system validity. For these tests the following parts of the SoC algorithm were implemented in the real-time SoC evaluation system: the newly developed SoC=f(EMF) model (EMF_m) as a function of temperature and hysteresis (see Eq. (7.14)), V_{pm}, the new SoC_{lm} as a function of the C-rate current, SoC_s and temperature (see Eq. (7.16)) and the maximum capacity Q_{max} adaptation algorithm (see Fig. 7.3 and Eq. (7.1)).

To allow for the hysteresis effect in the EMF, two sets of parameters were stored in the real-time SoC evaluation system for the EMF_m (see also Table 7.2). When the system was in the charge state before transition, the parameters corresponding to the charge state were used as input for the EMF_m, while the parameters corresponding to the discharge were used as input for the EMF_m when the previous state was discharge. Tests under conditions similar to those presented in Table 7.1 were carried out with the improved SoC algorithm to verify the SoC and remaining run-time accuracy. Table 7.4 shows the results (see also Table 7.1 for comparison).

Table 7.4. Results obtained with the improved real-time SoC evaluation system using fresh batteries.

C-rate current	T [°C]	SoC_{st} [%]	SoC_l [%]	t_{rstp} [min]	t_{rstm} [min]	t_{re} [min]	t_{rre} [%]	SoC_{end} [%]
0.10	5	97.4	3.7	601.6	599.8	1.8	0.3	4.2
		61.8	3.1	376.9	370.0	6.9	1.9	4.2
	25	100.0	2.4	626.6	632.6	−6.0	−0.9	1.4
		38.2	1.7	234.3	234.3	0.0	0.0	1.7
	45	100.0	1.6	631.7	638.9	−7.2	−1.1	0.4
		51.5	1.3	322.3	324.3	−2.0	−0.6	1.6
0.25	5	97.3	5.0	237.0	234.7	2.3	1.0	5.9
		42.7	3.9	99.6	102.7	−3.1	−3.0	2.7
	25	100.0	3.2	248.6	251.1	−2.5	−1.0	2.1
		36.3	2.3	87.3	86.5	0.8	0.9	3.4
	45	100.0	2.2	251.2	254.7	−3.5	−1.4	0.8
		51.3	1.8	127.1	127.6	−0.5	−0.4	2.0
0.50	5	96.9	6.2	116.5	113.9	2.6	2.3	8.1
		41.8	5.0	47.3	46.9	0.4	0.9	5.4
	25	100.0	4.0	123.3	124.7	−1.4	−1.1	3.4
		36.2	3.1	42.6	42.3	0.3	0.7	3.4
	45	100.0	2.8	124.8	126.2	−1.4	−1.1	1.6
		51.4	2.4	62.9	62.9	0.0	0.0	2.4
0.75	5	96.5	6.4	77.1	77.2	−0.1	−0.1	5.9
		40.8	5.9	29.9	30.1	−0.2	−0.7	5.5
	25	100.0	4.6	81.7	82.7	−1.0	−1.2	3.2
		57.1	3.9	45.5	46.0	−0.5	−1.1	4.6
	45	100.0	3.3	82.8	83.7	−0.9	−1.1	2.1
		51.3	2.9	41.4	41.7	−0.3	−0.7	2.6
1.00	5	96.7	7.6	57.2	56.3	0.9	1.6	8.0
		39.7	6.8	21.1	20.8	0.3	1.4	7.4
	25	100.0	5.0	60.1	60.9	−0.8	−1.3	3.5
		36.2	4.6	20.3	20.2	0.1	0.5	4.7
	45	100.0	3.6	61.9	62.6	−0.7	−1.1	2.4
		51.1	3.4	30.6	31.0	−0.4	−1.3	2.7

One example from Table 7.4 will be further discussed below. An SoC_{st} of 41.8% and an SoC_l of 5% were calculated using the newly developed SoC=f(EMF) and SoC_l models at the beginning of the partial discharge step performed at 0.5 C-rate and 5°C. The SoC evaluation system consequently calculated a t_{rstp} of 47.3 minutes (see Eq. (7.2)), whereas the t_{rstm} amounted to 46.9 minutes. This implies an inaccuracy in remaining run-time prediction of 0.4 minute or 0.9%.

It follows from Table 7.4 that the improved SoC evaluation system yields a remaining run-time accuracy with an error of less than 3.1 minutes at t_{rstm} values lower than 100 minutes and an error of less than 3.0% at t_{rstm} values higher than 100 minutes. In the majority of the measured cases an error of less than 1 minute, or 1%, was however obtained. This leads to the conclusion that the improved SoC evaluation system offers more accurate remaining run-time indication.

Fig. 7.12 presents the t_{re} [min.] as a function of t_{rstp} [min.] to show how close the measured results are to the data presented in Table 7.4 (see also Fig. 7.5 for comparison).

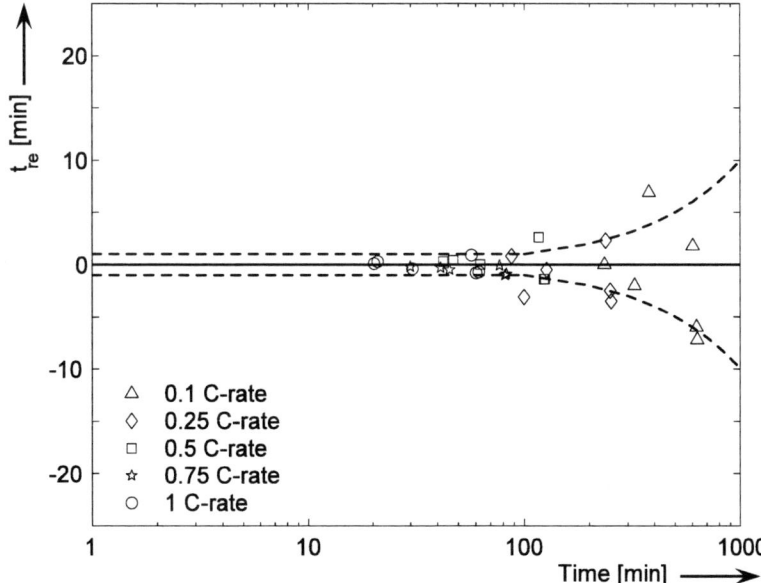

Fig. 7.12. The relative error in remaining run-time as a function of time in the improved SoC evaluation system tested at constant discharge C-rate currents.

A second set of tests were carried out to further validate the accuracy of the SoC evaluation system under an extended range of conditions. In these tests different switching C-rate currents were used for the discharging at different SoC_{st} values and different temperatures. The results of these tests verified the accuracy of the SoC_l model under different discharging conditions with due allowance for the battery's behaviour over time. The switching between the C-rate currents was each time effected at 3.4 V. This low voltage level was chosen because the accuracy of the remaining run-time prediction was expected to be poorer at lower voltage/SoC levels. The measurement results are summarised in Table 7.5 (see also Table 7.1). The current switched from the C-rate initial value to the C-rate final value when the battery reached the 3.4 V voltage level.

Table 7.5. Results obtained with the improved real-time SoC evaluation system using fresh batteries and switching discharge C-rate currents.

Initial C-rate current	Final C-rate current	T [°C]	SoC_{st} [%]	SoC_l [%]	t_{rstp} [min]	t_{re} [min]	t_{rre} [%]	SoC_{end} [%]
0.10	0.20		43.3	2.9	259.5	−3.2	−1.2	1.9
0.25	0.35		41.3	3.9	96.0	−1.4	−1.4	3.1
0.50	0.40	5°C	42.1	5.0	47.6	−0.6	−1.2	4.75
0.75	0.65		47.9	5.7	36.1	−0.2	−0.6	5.6
1.00	0.90		45.1	5.8	25.2	−1.7	−6.3	3.4
0.10	0.20		56.8	1.8	352.8	−4.3	−1.2	3.2
0.10	0.50		98.2	2.3	615.7	−1.5	−0.2	1.2
0.10	0.50		54.0	1.8	335.2	−2.3	−0.7	0.0
0.10	1.00		98.4	2.3	617.0	−0.8	−0.1	1.1
0.10	1.00		54.0	1.8	335.2	−1.5	−0.4	0.0
0.25	0.35		35.1	2.3	84.2	1.6	1.9	3.2
0.50	0.40	25°C	56.3	3.3	68.0	−1.7	−2.4	4.4
0.50	1.00		98.5	3.9	121.5	−0.6	−0.5	3.0
0.50	1.00		54.0	3.4	65.0	−1.1	−1.7	1.7
0.75	0.65		56.5	3.9	45.0	−0.9	−2.0	4.9
1.00	0.50		99.2	4.8	60.6	−2.1	−3.3	3.3
1.00	0.50		58.3	4.7	34.4	1.8	−5.5	6.1
1.00	0.90		35.2	4.5	19.7	0.5	2.6	5.2
0.10	0.20		51.5	1.2	323.0	−0.4	−0.1	1.4
0.25	0.35		51.6	1.8	127.9	−0.3	−0.2	1.9
0.50	0.40	45°C	51.2	2.4	62.7	−0.3	−0.5	2.2
0.75	0.65		51.6	2.9	41.7	0.0	0.0	2.9
1.00	0.90		51.4	3.4	30.8	−0.4	−1.3	2.8

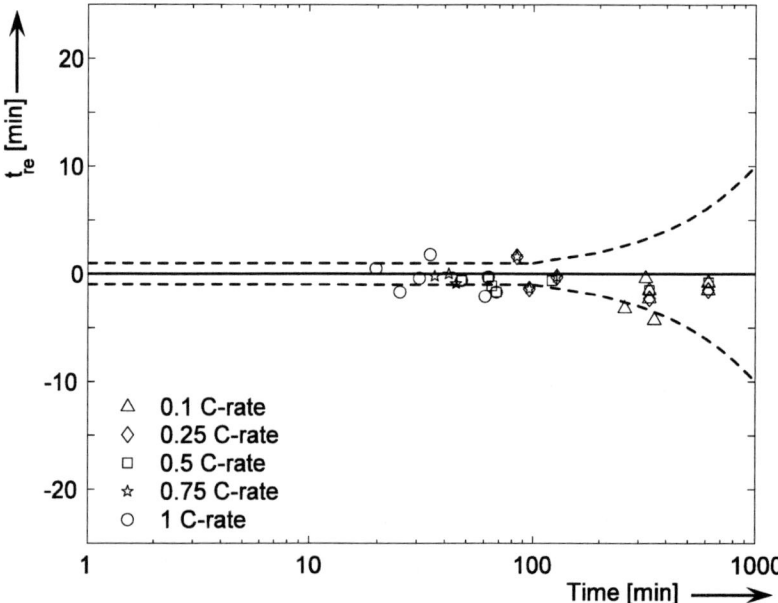

Fig. 7.13. The relative error in remaining run-time as a function of time in the improved SoC evaluation system tested at switching discharge C-rate currents.

It follows from Table 7.5 that the SoC evaluation system calculations lead to an error of –0.8 minute or –0.1% in the case of full discharge effected with a switching C-rate current from 0.1 to 1 C. This means that the improved SoC evaluation system yields a maximum error of 2.1 minutes, corresponding to 1.7%, in remaining run-time prediction. In the majority of the cases presented in Table 7.5 the error in remaining run-time prediction is however less than 1 minute, or 1%.

The relative error in remaining run-time as a function of t_{rstp} is indicated in Fig. 7.13 to show how close the results presented in Table 7.5 are to the actual data (see also Figs. 7.5 and 7.12 for comparison).

7.5.5 Uncertainty analysis

Fig. 7.14 shows a diagram of the sources of error in the newly developed real-time SoC evaluation system.

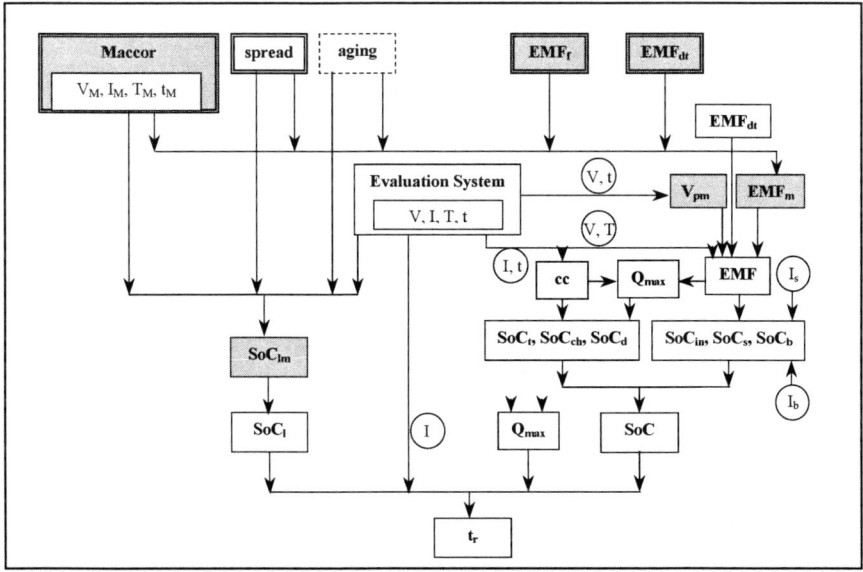

Fig. 7.14. Schematic representation of the sources of error in the improved real-time SoC evaluation system (see also Fig. 7.6 for comparison).

The new SoC_l model contains fewer sources of error in the remaining run-time calculation. In this case SoC_l is calculated by means of one function. Accurate results were obtained with this model at different temperatures and C-rate currents (see Figs. 7.10 and 7.11). Another important improvement is the new SoC=f(EMF) model that also includes the temperature dependence and the charge/discharge hysteresis. Accurate results were obtained with this model for both charge and discharge EMFs (see Figs. 7.8 and 7.9). These two important improvements, identification of the sources of error and the subsequent uncertainty analysis, greatly improve the real-time SoC system's accuracy in the case of fresh batteries.

7.6 Comparison with Texas Instruments' bq26500 SoC indication IC

The SoC evaluation system presented in this chapter was tested together with the bq26500 SoC IC developed by Texas Instruments [17] for the purpose of comparison. Regarding our best knowledge at the time of performing these tests the bq26500 SoC monitor was one of the most advanced SoC indicators commercially available.

7.6.1 The bq26500 SoC indicator

As mentioned in chapter 2 the bq26500 runs a book-keeping algorithm, *i.e.* Coulomb-counting algorithm combined with compensation for discharge C-rate current, temperature and self-discharge rate to calculate SoC and the remaining run-time available at one C-rate discharge current. In order to allow for the aging effect (see Fig. 6.1) the battery capacity is learned in the course of a discharge cycle from full to empty.

The bq26500 SoC system contains several programmable registers accessible through a high-speed single-wire interface (HDQ), in which measured data, such as the rated battery capacity, compensation factors for the discharge C-rate current, temperature and self-discharge rate or the V_{EoD} level, must be stored beforehand. The system chooses the compensation factors on the basis of voltage, current and temperature measurements. For a better understanding of the bq26500 system functionality, Fig. 7.15 gives the SoC and remaining run-time calculation diagram during charging and discharging.

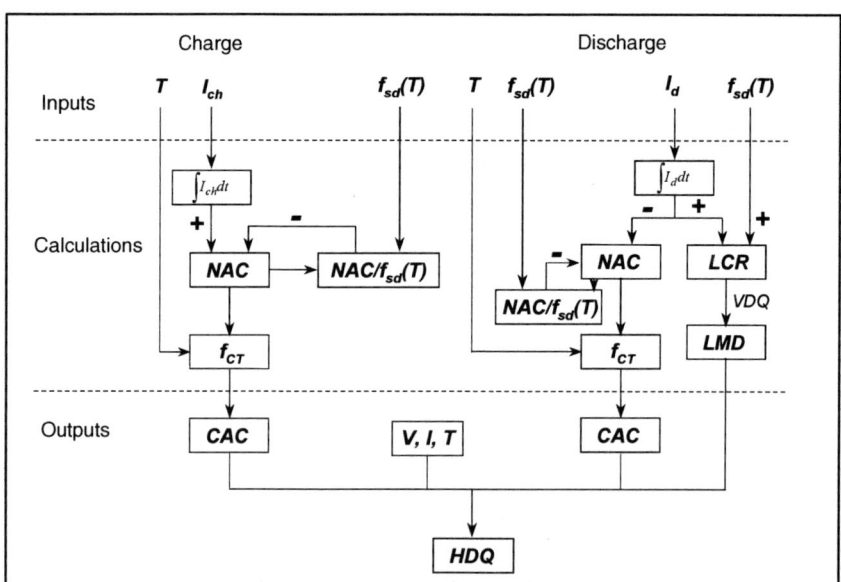

Fig. 7.15. Diagram of the SoC and remaining run-time calculation performed by the bq26500 (modified from [17]).

As can be seen in Fig. 7.15, four registers are important in SoC calculation in the bq26500:

NAC: Nominal Available Charge;
CAC: Compensated Available Charge;
LCR: Learning Count Register;
LMD: Last Measured Discharge.

All four registers count in 3µVh units. To obtain a mAh unit the measured value must be divided by the value of the sense resistor, *i.e.* 20 mΩ. The NAC and CAC registers are the actual results of the capacity estimation. Basically, the NAC is a Coulomb counter compensated only for self-discharge (see Fig. 7.15). The user must set the self-discharge estimation rate (f_{sd}) at 25°C. This rate is used to estimate the self-discharge capacity loss in one day when the battery is not being charged. The self-discharge rate is compensated for temperature by doubling the programmed rate for every 10°C increase or halving the programmed rate for every 10°C decrease. The CAC register reflects the capacity that is available under the current discharge conditions or the charge that is stored under the current charge conditions. It is obtained from the NAC register by multiplying its contents by a factor (f_{CT}) smaller than or equal to one. Therefore, the value stored in the CAC register is always equal to or smaller than the contents of the NAC register. The f_{CT} factor must be calculated on the basis of discharge cycle measurements executed at different C-rate currents and temperatures. The multiplication factors will change and CAC will acquire a different value as soon as the conditions (I and/or T) change.

The bq26500 measures a battery's capacity under charge/discharge conditions and updates the last measured discharge (LMD) register with the latest measured value. The bq26500 retains the learned LMD value unless a full reset occurs. By measuring the capacity during a discharge cycle from full to a stored End-of-Discharge Voltage level (EDV_1) without any disqualifying events, the bq26500 learns a battery's maximum capacity. The EDV_1 must be set to the threshold at which the remaining battery capacity is expected to be 6.25% of the maximum capacity. The bq26500 does not learn the capacity between EDV_1 and V_{EoD} levels, but assumes that the capacity is 6.25% of the LMD. The learned count register (LCR) has a value of LMD*6.25% added to the measured discharge from full to EDV_1, so for the system's SoC calculation accuracy it is important to set the EDV_1 level to a value close to this 6.25%. NAC is adjusted to LMD/16 during a discharge (unless NAC already has a smaller value) when EDV_1 is reached. If NAC reaches LMD/16 before the EDV_1 level is reached, NAC is held at the LMD/16 value until the EDV_1 level is reached. So NAC is synchronised to the 6.25% capacity level at the EDV_1 threshold.

The learning discharge cycles are disqualified by several conditions, which set a valid discharge flag (VDQ). If a learning cycle occurs with a significant reduction in learned capacity, the new LMD value will be restricted to a maximum LMD reduction during any single learning discharge of LMD/8. The maximum capacity will not be updated under the following conditions:

a) low temperature: temperatures less than or equal to a certain programmed value, *e.g.* lower than 25°C, when the EDV_1 threshold voltage is reached;

b) light load: average currents less than or equal to 2 times the standby current when the EDV_1 threshold voltage is reached;

c) fast voltage drop: $\Delta V \leq (EDV_1 - 256 \text{ mV})$ when EDV_1 is reached;

d) excessive charging: if the cumulative charge is greater than 255 mAh during a learning discharge cycle;

e) reset: VDQ is cleared on reset;

f) excessive self-discharge: if NAC reduction from self-discharge estimate exceeds 12.48%;

g) self-discharge at termination of learning cycle, *e.g.* self-discharge reduces NAC until NAC $\leq$ LMD/16.

The bq26500 can also calculate the remaining run-time in minutes for one discharge current value. The user must programme the discharge current value. The remaining run-time is further calculated by dividing the value of the CAC register by the value of the stored discharge current value.

7.6.2 Comparison of the two SoC indicators

To enable comparison, a set of tests similar to those whose results are presented in Tables 7.1 and 7.4 were carried out using the bq26500. Table 7.6 summarises the results. The discharge C-rate current and the temperature T in [°C] at which the measurements were carried out are given in columns one and two. The SoC, the predicted remaining run-time indicated at the start of the test and the remaining run-time measured during discharging to the 3V level are given in columns three, four and five, respectively (as in Table 7.1). The SoC calculated at the EDV_1 level, SoC_{EDV1}, is given in column six. As can be seen in Table 7.6, in some situations the bq26500 indicated an SoC value differing from 6.25% at the EDV_1 level. Columns seven and eight denote the predicted remaining run-time at the SoC_{EDV1} level, t_{rp} and the measured remaining run-time from the SoC_{EDV1} level, t_{rm}. The remaining run-time indicated at the end of the experiment, t_{re}, and the error in the remaining run-time, t_{rer}, calculated as the difference between the measured remaining run-time from the SoC_{EDV1} level, t_{rm}, and the predicted remaining run-time at the SoC_{EDV1} level t_{rp}, are given in columns nine and ten. Consequently, when $t_{rm} > t_{rp}$ the bq26500 will make a pessimistic estimation and when $t_{rm} < t_{rp}$ the bq26500 will make an optimistic estimation. For consistency in the results of the comparison the predicted remaining run-time was calculated by dividing the value of the CAC register by the actual rate discharge current AR_d [µV]. In this case the value of AR_d was calculated as follows

$$AR_d = \frac{I_d[mA] * R[m\Omega]}{LSB} \qquad (7.17)$$

where the *LSB* is 3µ.

Table 7.6. Full and partial discharge at constant C-rate currents and T using Texas Instruments' bq26500 IC*.

C-rate	T [°C]	SoC$_{st}$ [%]	t$_{rstp}$ [min]	t$_{rstm}$ [min]	SoC$_{EDV1}$ [%]	t$_{rp}$ [min]	t$_{rm}$ [min]	t$_{re}$ [min]	t$_{rer}$ [min]
0.10	5	93.9	206.3	491.5	6.2	18.4	275.4	0.0	257.0
		0.0	2.3	4.2	–	–	–	–	1.9
		0.0	2.3	8.9	–	–	–	–	6.6
	25	100.0	421.0	560.6	6.2	26.0	87.6	0.0	61.6
		12.8	37.8	34.9	6.2	18.4	34.8	0.0	16.4
		6.2	18.4	24.6	6.2	18.4	24.6	0.0	6.2
	45	11.9+	74.7	551.1	–	–	–	–	476.4
		32.0	134.6	132.2	6.2	26.3	38.7	0.0	12.4
		16.4	69.1	67.8	6.2	26.3	36.1	0.0	9.8
0.25	5	70.8	55.1	184.9	6.2	7.4	148.9	0.0	141.5
		0.0	2.7	0.4	–	–	–	–	–2.3
		1.6	1.8	1.3	–	–	–	–	–0.5
	25	100.0	201.5	219.6	7.0	12.6	93.2	0.0	80.6
		13.0	15.3	12.5	6.2	7.4	12.4	0.0	5.0
		6.2	7.4	8.4	6.2	7.4	8.4	0.0	1.0
	45	100.0	251.3	223.6	6.2	15.7	29.7	0.0	14.0
		31.3	52.8	51.1	6.2	10.5	32.4	0.0	21.9
		15.5	26.2	24.9	6.2	10.5	23.4	0.0	12.9
0.50	5	63.1	23.0	86.5	6.2	3.7	85.8	0.0	82.1
		0.6	0.3	0.1	–	–	–	–	–0.2
		0.0	0.0	0.1	–	–	–	–	0.1
	25	100.0	88.2	62.6	7.1	5.5	42.0	0.0	36.5
		12.7	7.5	4.7	6.0	3.6	4.5	0.0	0.9
		6.2	3.7	3.0	–	–	–	–	–0.7
	45	100.0	125.7	111.5	6.2	7.9	34.9	0.0	27.0
		31.0	26.1	24.5	6.2	5.3	24.4	0.0	19.1
		15.0	12.7	11.5	6.2	5.3	11.4	0.0	6.1
0.75	5	84.3	26.5	58.6	6.2	2.5	58.4	0.0	55.9
		0.0	0.0	0.1	0.0	0.0	0.1	0.0	0.1
		1.6	0.9	0.1	1.6	0.9	0.1	0.8	–0.8
	25	100.0	51.4	33.3	6.9	3.1	18.8	0.0	15.7
		12.5	4.9	1.5	6.2	2.5	1.4	0.0	–1.1
		6.2	2.5	0.6	6.2	2.5	0.6	0.0	–1.9
	45	100.0	73.3	73.8	6.1	4.5	36.8	0.0	32.3
		30.7	17.2	15.5	6.1	3.4	15.5	0.0	12.1
		14.7	8.3	7.0	6.1	3.4	6.9	0.0	3.5
1.00	5	91.6	22.9	43.9	6.0	1.8	43.8	0.0	42.0
		0.4	0.1	0.3	–	–	–	–	0.2
		4.2	1.2	0.2	–	–	–	–	–1.0
	25	100.0	33.8	49.6	7.1	2.1	47.5	0.0	45.4
		12.4	3.7	0.2	–	–	–	–	–3.5
		6.2	1.8	0.2	–	–	–	–	–1.6
	45	100.0	48.1	54.8	7.1	3.0	33.9	0.0	30.9
		30.4	12.8	11.2	6.1	2.6	11.0	0.0	8.4
		14.4	6.1	4.8	6.1	2.6	4.6	0.0	2.0

*The TCOMP register has been programmed for these tests to the default value of 0x7C [18]
+ Value indicated after a system reset and a full charge cycle.

Two examples from Table 7.6 will be further explained. In the case of 0.5 C-rate current and 25°C, SoC_{st} was 12.7% and the battery voltage was higher than the defined EoD_1 level. On the basis of these measurements the bq26500 SoC system calculated a t_{rstp} of 7.5 minutes, whereas the value of the t_{rstm} was 4.7 minutes. After 0.2 minute's discharge time the battery voltage reached the EoD_1 level and the bq26500 system set the SoC to a value of 6%. Consequently, a new value of 3.6 minutes was calculated for the remaining run-time, resulting in an 0.9 minute inaccuracy in the remaining run-time. This inaccuracy was calculated with due allowance for the remaining run-time value set at the EoD_1 level.

In the second example the battery was fully charged by means of the CCCV method after a system reset. It should be noted that in the case of a system reset the bq26500 SoC system also resets the SoC value to 0%. Consequently, even when a battery is at a high SoC level, a 0% SoC value will be indicated. In this experiment the bq26500 SoC system indicated an SoC of 11.9% at the end of the charging, whereas the actual SoC was 100%. This high inaccuracy in the SoC value is attributable to the high battery SoC level at the beginning of charging and the 0% SoC considered by the bq26500 system at this level. The battery was then discharged to the 3 V level at 0.1 C-rate current and 45°C. At the beginning of discharging a remaining run-time of 74.7 minutes was predicted whereas t_{rstm} was 551.7 minutes.

It follows from Table 7.6 that when tested under an extended range of conditions the bq26500 system consistently made very pessimistic remaining run-time estimations. t_{rm} was consistently larger than t_{rp}. This means that a user of that system will recharge a battery more often, leading to more wear-out. The inaccuracy in the bq26500 measurements is mainly attributable to inaccurate SoC calculation after a system reset and at the EDV_1 level. The battery maximum capacity adaptation algorithm will consequently not work accurately either. Due to the first factor the bq26500 will indicate 0% SoC when it is started after a reset, even if a battery is completely full. It should be noted that this is a common inaccuracy in book-keeping systems that do not calculate an initial SoC value. Due to the second factor the system will indicate 6.2% SoC on the basis of a voltage level measurement under any load condition. However, as shown in chapter 2, the battery voltage depends on many factors, including the discharge current, temperature, aging and battery chemistry. SoC indication based on voltage measurements under load conditions is therefore very inaccurate (see chapter 2).

7.7 Conclusions

A new SoC algorithm has been presented in this chapter. The algorithm combines the advantages of battery Electro-Motive Force (EMF) and overpotential with the voltage-relaxation predictive method and the maximum capacity adaptive method. To test the SoC and the remaining run-time accuracy the new SoC algorithm was implemented in a real-time SoC evaluation system

Measurement results obtained with the SoC evaluation system using fresh US18500G3 Li-ion batteries under an extended range of conditions have been presented. In these results the error in remaining run-time was each time less than 20.3 minutes or 23.1%. However, these results do not meet the goal of this work, *i.e.* prediction of the remaining run-time of a Li-ion battery with an uncertainty of

1 minute or less when t_{rstm} has a value lower than 100 minutes and an indication error of 1% or less when t_{rstm} has a value higher than 100 minutes under all realistic user conditions. To enable more accurate SoC and remaining run-time calculation the sources of error in the SoC evaluation system were identified. This led to the conclusion that accurate modelling of the EMF=f(SoC) relationship and the overpotential are not enough to ensure a high level of accuracy in remaining run-time, too. For this reason new methods for modelling the SoC=f(EMF) relationship and predicting a battery's State-of-Charge-left were developed.

In these new methods no mathematical inversion is needed to determine the SoC during equilibrium, and no battery overpotential calculation, voltage measurement or EMF model calculations under load conditions are needed for the SoC_l calculation, as in the prior-art remaining run-time prediction method. New measurements were performed with the improved SoC evaluation system under an extended range of conditions. They showed that, with predictions with an error of less than 1 minute or 1% in most cases, the new SoC evaluation system greatly improves remaining run-time prediction accuracy.

The SoC evaluation system presented in this chapter was compared with Texas Instruments' book-keeping bq26500 SoC IC. This comparison showed that the newly developed SoC evaluation system performs much better than the bq26500 SoC system under all the tested conditions. It should be noted that at the time of publication other newly developed SoC indicators are commercially available [19]-[21].

7.8 References

[1] H.J. Bergveld, W.S. Kruijt, P.H.L. Notten, Battery Management Systems, Design by Modelling, Philips Research Book Series, **1**, Kluwer Academic Publishers, Boston (2002)

[2] V. Pop, H.J. Bergveld, P.H.L. Notten, P.P.L. Regtien, State-of-Charge Indication in Portable Applications, IEEE International Symposium on Industrial Electronics, **3**, 1007-1012 (2005)

[3] V. Pop, H.J. Bergveld, P.H.L. Notten, P.P.L. Regtien, A real-time evaluation system for a State-of-Charge indication algorithm, Proceedings of the Joint International IMEKO TC1+ TC7 Symposium, **1**, 104-107 (2005)

[4] V. Pop, H.J. Bergveld, P.H.L. Notten, P.P.L. Regtien, Smart and Accurate State-of-Charge Indication in Portable Applications, IEEE Power Electronics and Drive Systems, **1**, 262-267 (2005)

[5] H. J. Bergveld, H. Feil, J.R.G.C.M. Van Beek, Method of predicting the state of charge as well as the use time left of a rechargeable battery, US Patent 6,515,453, filed November 30 (2000)

[6] J.H. Aylor, A. Thieme, B.W. Johnson, A battery State-of-Charge Indicator for Electric Wheelchairs, IEEE Trans. on Industrial Electronics, **39**, 398-409 (1992)

[7] T. Kikuoka, H. Yamamoto, N. Sasaki, K. Wakui, K. Murakami, K. Ohnishi, G. Kawamura, H. Noguchi, F. Ukigaya, System for measuring state of charge of storage battery, US Patent 4,377,787, filed August 8 (1980)

[8] G.R. Seyfang, Battery state of charge indicator, US Patent 4,949,046, filed June 21 (1988)

[9] M.W. Verbrugge, E.D. Tate Jr., S.D. Sarbacker, B.J. Koch, Quasi-adaptive method for determining a battery's state of charge, US Patent 6,359,419, filed December 27 (2000)
[10] H. J. Bergveld, V. Pop and P.H.L. Notten, Method of estimating the State-of-Charge and of the use time left of a rechargeable battery, and apparatus for executing such a method, Patent WO2005085889, filed February 23 (2005)
[11] V. Pop, H.J. Bergveld, P.H.L. Notten, P.P.L. Regtien, State-of-the-art of state-of-charge determination, Measurement Science and Technology Journal, **16**, R93–R110 (2005)
[12] V. Pop, H.J. Bergveld, P.H.L. Notten, P.P.L. Regtien, D. Danilov, J.H.G. Op het Veld, Modelling battery behaviour for accurate State-of-Charge indication, J. Electrochem. Soc., **153**, A2013–A2022 (2006)
[13] H.J. Bergveld, V. Pop and P.H.L. Notten, Apparatus and method for determination of the state-of-charge of a battery when the battery is not in equilibrium, Patent PH006795EP1, filed October 30 (2006)
[14] National Instruments Corporation, NI 6034E/ 6035E/6036E User Manual, www.ni.com
[15] P.P.L. Regtien, F. van der Heijden, M.J. Korsten, W. Olthuis, Measurement Science for Engineers, Kogan Page Science Publisher, London, (2004)
[16] Maccor Inc., Help system, Version 2.5.4
[17] Texas Instruments, Single-Cell Li-ion and Li-Pol Battery Gas Gauge IC For Handheld Applications (bq Junior Family), Doc. I.D. SLUS567A (2003)
[18] Texas Instruments, Determining the Values to Program the bq26500/bq26501 EEPROM, Doc. I.D. SLUA317 (2004)
[19] Texas Instruments, SBS 1.1-Compliant gas gauge enabled with impedance track technology for use with the bq29330 (bq 20z70-V110), Doc. I.D. SLUS742 (2006)
[20] Texas Instruments, SBS 1.1-Compliant gas gauge enabled with impedance track technology for use with the bq29330A (bq 20z80-V102), Doc. I.D. SLUS681A (2006)
[21] Texas Instruments, SBS 1.1-Compliant gas gauge enabled with impedance track technology for use with the bq29330 (bq 20z90-V110), Doc. I.D. SLUS743 (2006)

Chapter 8
Universal State-of-Charge indication for battery-powered applications

The battery EMF and overpotential were combined with the voltage-relaxation predictive method and the maximum capacity adaptive method in a new SoC evaluation system in chapter 7. Accurate results have been obtained with the SoC evaluation system using fresh batteries under an extended range of conditions. However, the SoC=f(EMF), overpotential and SoC_l models include a variety of parameters that change during cycling of the battery. For a more accurate determination of the SoC when a battery ages, innovative adaptive systems were developed in chapter 6.

Implementation aspects and results obtained with the overpotential adaptive system will be presented in section 8.2. The focus in section 8.3 will be on a new adaptive system that combines the SoC=f(EMF) adaptive model under equilibrium conditions with an SoC_l adaptive model under load conditions. Implementation aspects of the adaptive system in the SoC evaluation system will also be presented. Section 8.4 will present measurement results obtained with the adaptive SoC evaluation system. An uncertainty analysis will be discussed in section 8.5. The dream of the last 70 years of research in the field of SoC has been to design a universal SoC system that will adapt to any type of battery on line, without user intervention. Results obtained with the adaptive SoC evaluation system for another type of battery will be presented in section 8.6. These results prove that the developed SoC evaluation system is capable of adapting to batteries with different chemistries and of offering accurate, universal SoC indication. Designers will also be interested in the implementation requirements of the mathematical functions in a practical application. A possible implementation of the SoC evaluation system on a mobile phone platform will be presented in section 8.7. The usability of the SoC algorithm in a newly developed ultra-fast recharging algorithm will also be presented in this section. Finally, section 8.8 will present concluding remarks.

8.1 Introduction

The real-time SoC evaluation system combining an on-line predictive algorithm with the Q_{max} and overpotential adaptive algorithms presented in chapters 6 and 7 was tested using aged batteries. The goal of the SoC evaluation system is to predict the remaining run-time of any Li-ion battery with an uncertainty of 1 minute or less when t_{rst} has a value lower than 100 minutes and an indication error of 1% or less when t_{rst} has a value higher than 100 minutes under all realistic user conditions, including a wide variety of load currents and over a wide temperature range.

8.2 Implementation aspects of the overpotential adaptive system

The overpotential adaptive method presented in chapter 6 combines the measured $\eta_{ch}^a / \eta_{ch}^f$ ratio with the overpotential symmetry phenomenon. A diagram illustrating the overpotential adaptation mechanism is shown in figure 8.1.

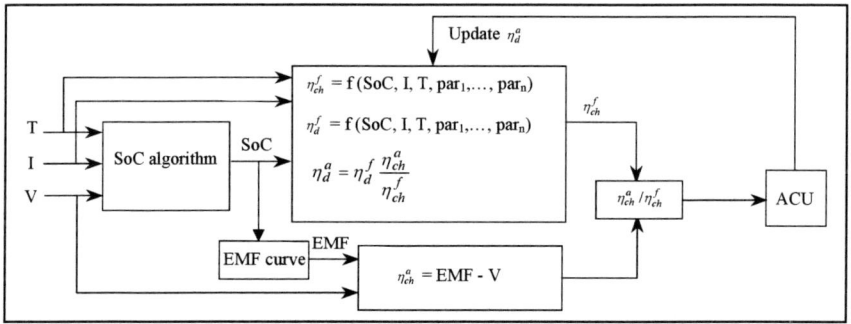

Fig. 8.1. Diagram of the overpotential adaptation mechanism.

The SoC is determined on the basis of a starting SoC value when entering charge mode by adding the accumulated charge obtained by integrating the charge current. The latest value of Q_{max} is used to obtain the SoC value on a percentage scale. Each time a new set of the battery variables V, I and T is measured, the SoC evaluation system measures a new charge battery overpotential. So, the ratio $\eta_{ch}^a / \eta_{ch}^f$ is determined and fed to an Adaptive Control Unit (ACU). On the basis of the new value of the ratio in relation to earlier ratio values the ACU decides whether to update η_d^a. This process is repeated an arbitrary number of times in the CC charging mode.

A set of tests was carried out using a US18500G3 Li-ion battery and the presented overpotential adaptation mechanism in order to test the system's accuracy. By the time of testing, the battery had undergone approximately 600 discharge/charge cycles. The tests were carried out using full charge/discharge cycles at different constant currents and 25°C. A maximum capacity value of 1176 mAh and overpotential parameters values obtained for a fresh battery (see Table 5.1) were stored in the SoC evaluation system at the beginning of the tests. During the charge cycles the battery was each time fully charged to 4.2 V with the normal CCCV charging method at 0.5 C-rate current in the CC mode. Each period of charging was followed by a rest period of 0.5 hour. After this rest period a discharge step at 0.5 C-rate current was applied until the battery voltage reached 3 V. This procedure was repeated using a 0.75 C-rate discharging current. During the charge cycle a new maximum capacity value of 1108 mAh and a value of 1.4 for the $\eta_{ch}^a / \eta_{ch}^f$ ratio were obtained [1]. The measurements described above were repeated with the newly determined parameter values using different constant discharge currents (0.1, 0.25, 0.5, 0.75 and 1 C-rate).

Table 8.1 summarises the experimental results. A distinction has been made between results obtained without an adaptive system and the results obtained using the newly developed adaptive system. The discharge C-rate current at which these tests were carried out is given in column 1. The SoC indicated at the start, SoC_{st}, and the SoC calculated by means of the overpotential and EMF models, expressed as SoC_l, are given in columns two and three. Columns four, five and six denote the remaining run-time in minutes at the start of the experiment, the error in the remaining run-time at the end of the experiment and the relative error in the remaining run-time.

Table 8.1 Results obtained with the real-time SoC evaluation system.

C-rate current	SoC_{st} [%]	SoC_l [%]	t_{rst} [min]	t_{re} [min]	t_{rre} [%]
Measurement results obtained without adaptation					
0.50	98.3	3.4	121.8	7.1	6.2
0.75	98.3	4.1	80.6	5.6	7.5
Measurement results obtained with adaptation					
0.10	98.2	2.7	577.2	−18.0	−3.0
0.25	98.2	3.2	229.7	−3.6	−1.5
0.50	98.2	4.0	113.9	−0.5	−0.4
0.75	98.2	4.9	75.2	0.3	0.4
1.00	98.2	6.0	55.7	0.6	1.1

It follows from Table 8.1 that when the SoC evaluation system was used without an adaptive system, relative errors of 6.2% and 7.5% in remaining run-time were obtained at discharge currents of 0.5 and 0.75 C-rate. Under these conditions the SoC indicator made very optimistic estimations. For instance, at the start of the discharge step at 0.5 C-rate the system indicated 121.8 minutes remaining run-time. After 114.7 minutes the battery reached the level of 3 V. This means that t_{rre} calculated by Eq. (7.4) was 6.2%. In a second example the overpotential and maximum capacity adaptive methods were used to adapt the model parameters. It follows from Table 8.1 that, after adaptation, t_{rre} values of −0.4% and 0.4% were obtained for the 0.5 and 0.75 C-rate discharge currents, respectively. Under these conditions the SoC indicator made slightly pessimistic and optimistic estimations, respectively. For instance, at the start of the discharge step at 0.5 C-rate the system indicated 113.9 minutes remaining run-time. After 114.3 minutes the battery reached the level of 3 V, implying a t_{rre} value of −0.4%. It can be concluded that the adaptive SoC system yielded an error in remaining run-time of less than 0.6 minute, or 3%, in all cases. However, as discussed in chapter 7, accurate modelling of the EMF=f(SoC) relationship and overpotential is not enough to obtain a high level of accuracy in remaining run-time prediction when a battery is partially discharged. Therefore, new functions for the SoC=f(EMF) and State-of-Charge-left (SoC_l) were developed in chapter 7. These functions led to a much higher level of accuracy in remaining run-time prediction in the case of fresh batteries.

8.3 SoC=f(EMF) and SoC_l adaptive system

To ensure accurate SoC and t_r calculation while a battery ages, variations in the EMF and State-of-Charge-left model parameters need to be considered. Chapter 6 described a new adaptive method for determining a battery's EMF when the battery is in equilibrium. In this section a new adaptive method for determining a battery's SoC_l when the battery is under load conditions will be presented. This adaptive system was developed on the basis of phenomena discovered during the measurement analysis.

In this method the SoC_l model parameters are adapted to take a battery's aging process into account. Each time a battery is discharged to the V_{EoD} level an EMF value is predicted on the basis of the first few minutes of the relaxation process by means of the voltage-relaxation model presented in chapter 4 [2]. A measurement example is shown in Fig. 8.2, which illustrates what happens to a battery's Open-Circuit Voltage (OCV) after a discharge step from 100% SoC at 0.1 C-rate and 25°C to the V_{EoD} level of 3 V. To enable comparison, the voltage V_p predicted on the basis of the OCV measured in the first 15 minutes of the relaxation process, OCV_m, is also shown. The measurement was carried out using a 5.4% capacity loss battery (see chapter 6).

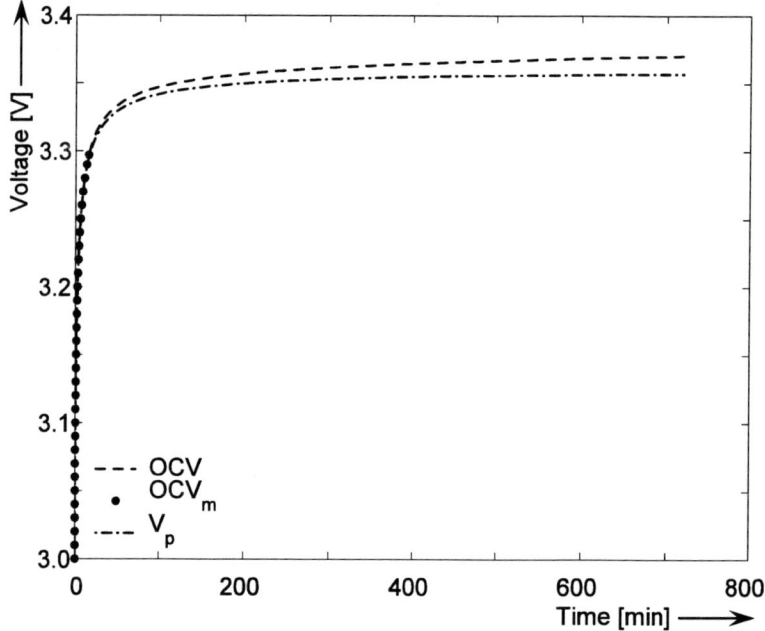

Fig. 8.2. The battery Open-Circuit Voltage and the voltage predicted after a discharge step from 100% SoC at 0.1 C-rate current and 25°C as a function of the relaxation time in [min.].

As can be seen in Fig. 8.2, after a complete discharge step, the value of the battery's OCV changes from 3.0 V immediately after the current interruption to about 3.37 V after 720 minutes. It follows from Fig. 8.2 that the voltage-prediction value based on the OCV measured in the first 15 minutes of the relaxation process is very close to the EMF value measured after 720 minutes. In this example the difference is only 13 mV. The predicted EMF value can be used as input for the adaptive SoC=f(EMF) model. In this case an SoC_l value under the applied measurement discharge condition can be estimated. In this example a –0.1% SoC inaccuracy is obtained by comparing the SoC values obtained by means of EMF and by means of V_p. So, each time a battery is discharged to the V_{EoD} level a new SoC_l value can be accurately determined on the basis of V_p and the adaptive SoC=f(EMF) relationship.

A general block diagram of how the SoC-EMF and SoC_I adaptive system could be implemented in the SoC indication system is given in Fig. 8.3.

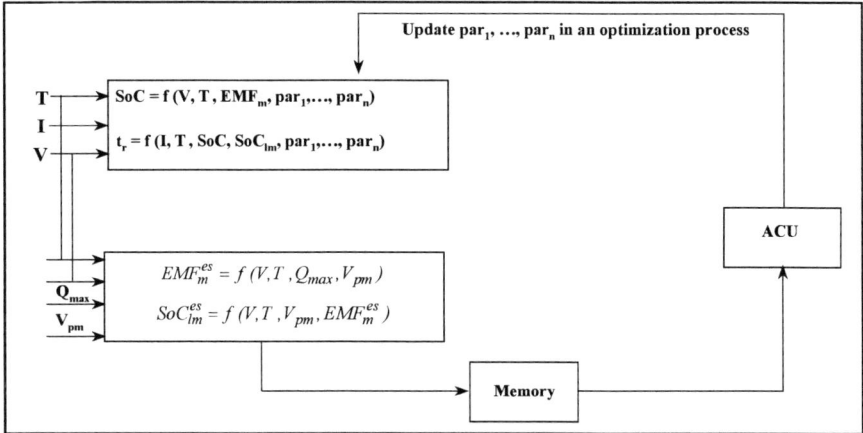

Fig. 8.3. Mechanism for updating parameters $par_1, \ldots, par_n$ in the SoC=f(EMF) and SoC_I models.

An SoC value is calculated by means of battery voltage and temperature measurements and the stored SoC=f(EMF) model (EMF_m) when the battery is in equilibrium. When current is flowing through the battery a remaining run-time value is calculated by means of current and temperature measurements and the SoC_I model (SoC_{Im}). The EMF_m and SoC_{Im} contain a set of parameters $par_1, \ldots, par_n$ that need to be updated when a battery ages to ensure more accurate battery SoC and remaining run-time calculations. After each current interruption a new set of battery variables V and T is measured and the SoC adaptive and predictive algorithm estimates new EMF (EMF_m^{es}) and SoC_I (SoC_{Im}^{es}) values. These estimated values are stored in a memory, *e.g.* EEPROM. This process is repeated an arbitrary number of times after current interruption. The estimated samples are fed to an Adaptive Unit that decides whether to update the parameter set $par_1, \ldots, par_n$ of the EMF_m and SoC_{Im} used for the SoC and remaining run-time calculation (see Fig. 8.3). Any optimisation algorithm can be used in the adaptive algorithm. Several examples can be found in the literature. Note that when the adaptive system is implemented as described in this chapter it will work at any values of V and T.

8.4 Results obtained with the adaptive SoC system using aged batteries

In order to verify the SoC and remaining run-time accuracy, two sets of tests were carried out using the adaptive SoC evaluation system at different constant and switching C-rate currents and temperatures. The SoC=f(EMF) adaptive model as a function of temperature and hysteresis and the SoC_I adaptive model as a function of the C-rate current, SoC_s and temperature were combined with the cc, V_{pm} and the Q_{max} adaptation algorithm to obtain a complete on-line adaptive system.

The adaptive real-time SoC evaluation system was first tested in a set of tests similar to those whose results were presented in table 7.4. A 5.4% capacity loss battery was used in all these tests (see chapter 6). New EMF_m and SoC_{lm} parameters were learned off-line by means of the adaptive methods presented in section 8.3. Partial battery discharging was also considered in these tests, to prove that accurate modelling and adaptation of the SoC=f(EMF) relationship and the SoC_{lm} lead to a high level of accuracy in remaining run-time prediction under an extended range of conditions. This is an advantage over the algorithm presented in sections 7.2 and 8.2, which yielded inaccurate remaining run-time results under partial discharge conditions. The results are presented in Table 8.2 (compare with Table 8.1).

Table 8.2 Results obtained with the adaptive SoC evaluation system using aged batteries.

C-rate current	T [°C]	SoC_{st} [%]	SoC_l [%]	t_{rstp} [min]	t_{rstm} [min]	t_{re} [min]	t_{rre} [%]	SoC_{end} [%]
0.10	5	94.8	7.1	516.6	497.3	19.3	3.9	3.9
		12.5	2.6	58.3	64.1	−5.8	−9.0	1.6
	25	99.5	2.9	569.1	576.8	−7.7	−1.3	1.5
		20.1	1.3	110.8	108.8	2.0	1.8	1.7
	45	100.0	1.8	578.5	585.6	−7.1	−1.2	0.6
		27.3	1.1	154.3	157.1	−2.8	−1.8	0.6
0.25	5	94.6	13.8	190.4	198.4	−8.0	−4.0	9.5
		6.8	3.8	7.1	8.4	−1.3	−15.5	3.2
	25	99.25	4.9	222.3	224.8	−2.5	−1.1	3.8
		20.1	2.3	41.9	41.1	0.8	1.9	2.6
	45	14.4	1.5	30.4	27.2	3.2	11.8	2.8
0.50	5	95.2	17.5	91.5	94.5	−3.0	−3.2	12.9
		12.4	7.0	6.4	7.1	−0.7	−9.9	6.3
	25	99.0	7.3	108.0	108.8	−0.8	−0.7	6.5
		20.2	3.8	19.3	19.4	−0.1	−0.5	3.8
	45	100.0	4.3	112.8	113.5	−0.7	−0.6	3.8
		27.0	2.7	28.6	29.3	−0.7	−2.4	2.1
0.75	5	95.2	23.9	56.0	58.0	−2.0	−3.4	17.5
		8.6	8.5	0.1	0.1	0.0	0.0	8.5
	25	99.0	9.4	70.4	70.5	−0.1	−0.1	8.8
		21.0	4.6	12.9	12.3	0.6	4.9	5.2
	45	100.0	5.5	74.2	74.1	0.1	0.1	5.7
		14.4	2.6	9.3	9.5	−0.2	−2.1	2.3
1.00	5	94.7	31.0	37.5	39.8	−2.3	−5.8	21.4
		12.7	12.7	0.0	0.0	0.0	0.0	12.6
	25	99.0	10.6	52.1	51.9	0.2	0.4	10.9
		21.0	5.2	9.3	8.6	0.7	8.1	6.5
	45	100.0	6.5	55.1	54.5	0.6	1.1	7.6
		14.6	4.9	5.7	5.0	0.7	14.0	6.1

It follows from Table 8.2 that at the beginning of the discharge step from 20.1% SoC at 0.10 C-rate and 25°C the system indicated 110.8 minutes remaining run-time. After 108.8 minutes the battery reached the level of 3 V. In this case an error of 2.0 minutes and a relative error of 1.8% in remaining run-time were calculated. Fig. 8.4 illustrates t_{re} [min.] as a function of t_{rstp} [min.] to show how close the measured data presented in Table 8.2 are to the actual data.

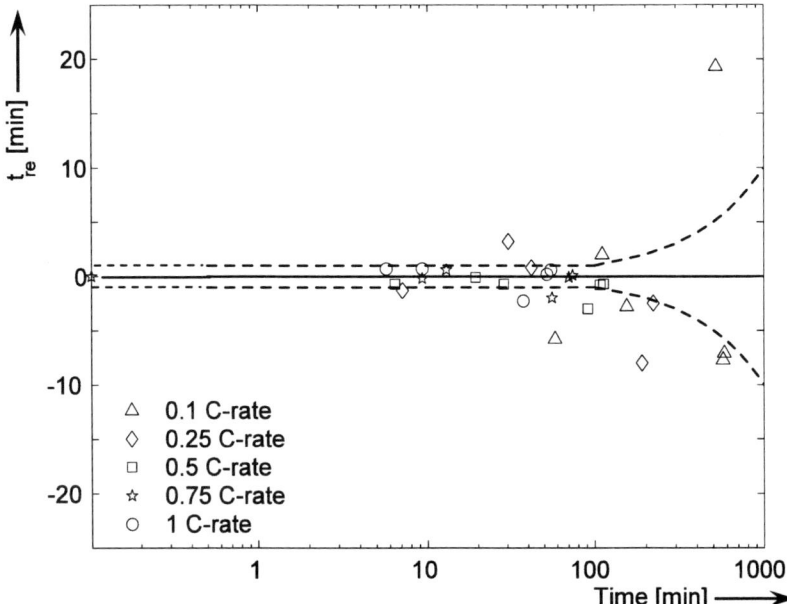

Fig. 8.4. Same as Fig. 7.12, only now showing results obtained with the adaptive SoC evaluation system using aged batteries.

It can be concluded that the newly developed adaptive system provides accurate remaining run-time prediction in the case of aged batteries. In our experiments a maximum error in remaining run-time accuracy of 5.8 minutes, or 4%, was each time obtained. In most cases the error in remaining run-time indication was however very close to the goal of 1 minute, or 1%.

The adaptive SoC evaluation system was subjected to a second set of tests, similar to those whose results were presented in Table 7.5, again using aged batteries. The results are summarised in Table 8.3 (see also Table 7.5). The current switched from the initial C-rate to the final C-rate when the battery reached the 3.4 V voltage level. The relative error in remaining run-time as a function of t_{rstp} is also indicated in Fig. 8.5, to show how close the measured data presented in Table 8.3 are to the actual data.

It follows from Table 8.3 and Fig. 8.5 that the improved SoC evaluation system yielded a maximum error in remaining run-time prediction of 1.6 minutes, or 2.3%. In most cases the error in remaining run-time indication was however less than 1 minute, or 1%.

Table 8.3 Same as Table 8.2, only now obtained at switching discharge currents.

C-rate current start	C-rate current end	T [°C]	SoC$_{st}$ [%]	SoC$_l$ [%]	t$_{rstp}$ [min]	t$_{rstm}$ [min]	t$_{re}$ [min]	t$_{rre}$ [%]	SoC$_{end}$ [%]
0.10	0.20		12.9	4.7	52.4	52.5	−0.1	−0.2	2.6
0.25	0.35		6.6	4.2	6.1	5.5	0.6	10.7	4.5
0.50	0.40	5°C	12.3	6.6	7.3	8.9	−1.6	−18.0	5.1
0.75	0.65		7.0	6.9	0.1	0.1	0.0	0.0	7.0
1.00	0.90		10.1	10.1	0.0	0.0	0.0	0.0	10.1
0.10	0.20		20.6	3.2	111.4	108.9	2.5	2.3	2.9
0.25	0.35		20.1	3.7	42.1	41.4	0.7	1.7	3.2
0.50	0.40	25°C	8.3	3.2	6.6	6.4	0.2	3.2	2.8
0.75	0.65		20.2	5.7	12.4	12.0	0.4	3.3	4.9
1.00	0.90		20.2	6.2	9.0	8.8	0.2	2.3	5.9
0.10	0.20		14.6	2.0	81.1	81.1	0.0	0.0	1.47
0.25	0.35		14.0	2.7	29.0	29.4	−0.4	−1.4	1.2
0.50	0.40	45°C	14.0	2.9	14.3	14.9	−0.6	−4.0	1.2
0.75	0.65		13.7	3.5	8.7	9.1	−0.4	−4.4	1.9
1.00	0.90		13.9	4.1	6.3	6.5	−0.2	−3.1	2.6

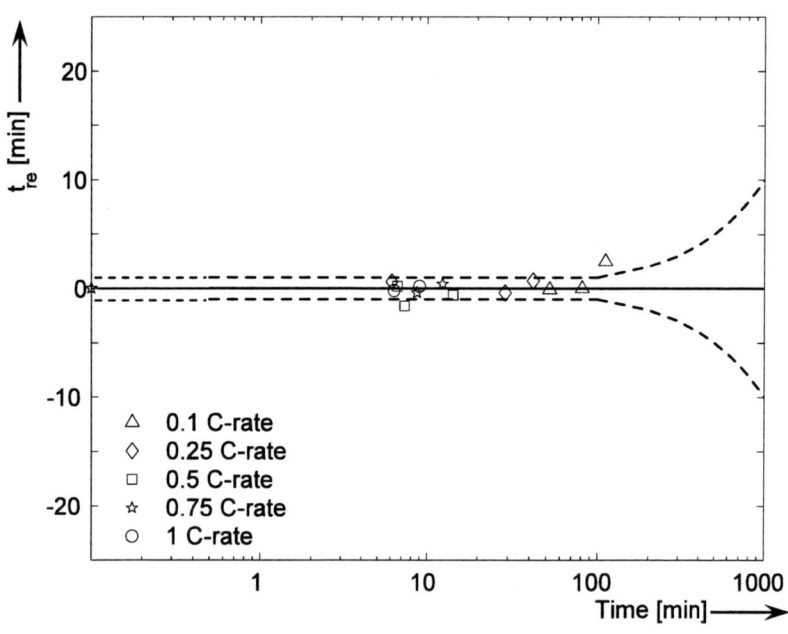

Fig. 8.5. Same as 8.4, only now obtained at switching discharge C-rate currents.

8.5 Uncertainty analysis

Fig. 8.6 presents a diagram showing the sources of error in the adaptive SoC evaluation system.

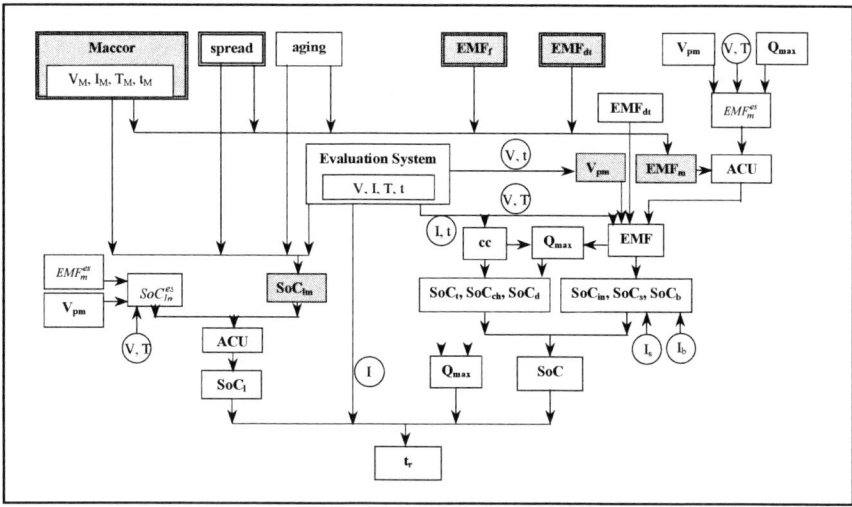

Fig. 8.6. Same as Fig. 7.14, only now for aged batteries (see also Fig. 8.3).

Fig. 8.6 contains two adaptive blocks for the EMF and SoC_l calculation that were not included in the fresh battery uncertainty analysis. Fig. 8.6 shows that the initial SoC=f(EMF) and SoC_l parameter values are updated on line via a decision ACU block (see also Fig. 8.3). This block should eliminate false adaptation situations that could adversely affect the remaining run-time indication accuracy. The adaptive system was developed to ensure accurate remaining run-time indication when a battery ages. It is based on the voltage-prediction model whose parameters are calculated on line and the maximum-capacity adaptive model. The system, which was developed in response to phenomena discovered during a thorough measurement analysis, greatly improves the real-time SoC system's accuracy in the case of aged batteries (see Tables 8.2 and 8.3).

8.6 Results obtained with other Li-based battery

The dream of the last 70 years of research in the field of SoC field has been to design an accurate SoC system that will adapt on-line to any type of battery without user intervention (see Chapter 2). A possible solution that may help realise this dream was given in Chapters 7 and 8 of this book. Accurate results have been obtained with this adaptive SoC algorithm under an extended range of conditions using fresh (see Chapter 7) and aged US18500G3 (Sony) batteries (see Chapter 8). To find out whether the adaptive system would yield similarly good results with other types of battery, further tests were carried out using a different type of Li battery. Table 8.4 presents the main characteristics of this battery. The main characteristics of the US18500G3 battery are also presented for comparison.

It follows from Table 8.4 that this different battery also has a different chemistry. This means that the EMF_m, SoC_{lm} and Q_{max} may show different behaviour, so use of the parameter values of the US18500G3 battery could lead to inaccurate SoC and t_r indication. This will be discussed below.

Table 8.4. US18500VR Li-ion battery characteristics.

Battery	Characteristics	
Chemical system	Ni + Mn	Co
Cell Type	VR	US18500G3
Cell diameter	18.4 mm max.	at most 18.4 mm
Cell Length	49.3 mm max.	at most 49.3 mm
Capacity (0.2 C-rate) Typical	1180 mAh (3.0 V cut off)	1180 mAh (3.0 V cut off)
Capacity (0.2 C-rate) Minimum	1100 mAh (3.0 V cut off)	1100 mAh (3.0 V cut off)
Cell weight	33 g	33 g

8.6.1 EMF and SoC$_l$ modelling results obtained for the Li-based battery

Fresh and fully activated batteries were used under each test condition. Since the employed commercial batteries were delivered with a specified SoC, the activation procedure started with Constant-Current discharging at 0.5 C-rate followed by a one-hour resting period. The batteries were then subjected to 3 standard CCCV and subsequent 0.5 C-rate discharge cycles, after which constant charge/discharge behaviour was observed. The batteries were subjected to standard (s) charging at a constant maximum current (I_s^{max}) at 1 C-rate in the CC-mode until the maximum charge voltage V_s^{max} of 4.2 V was reached in the subsequent CV-mode [3]. The charging currents were found to decrease in the CV-mode and charging was terminated at a predefined minimum current (I_s^{min}) of 0.05 C-rate, after which the batteries were assumed to be fully charged. After a resting period of one hour the batteries were discharged at 0.5 C-rate. Discharging was terminated when the cut-off cell voltage of 3.0 V was reached. A single activation cycle was followed by a one-hour resting period. The activation cycles were carried out at 25°C.

SoC=f(EMF) measurements and fitting

The maximum capacity and GITT measurement methods were combined with the voltage-prediction model for the EMF determination. A maximum capacity of 1220 mAh was first determined by means of the maximum capacity adaptation method described in chapter 7. For consistency the same EMF values of 3.0 and 4.175 V for the 0 and 100% SoC levels were defined for this method. Ten EMF predicted points with the corresponding SoCs were considered in the charge/discharge EMF determination. The voltage and time values measured during the 1-hour resting period were used as input for this prediction. The EMF points were fitted using the newly developed method in which the shape of the curve is also taken into consideration. The measurements were performed at 5, 25 and 45°C. The discharge EMF measured at 25°C is illustrated in Fig. 8.7. The EMF obtained for the US18500G3 battery is also shown for comparison.

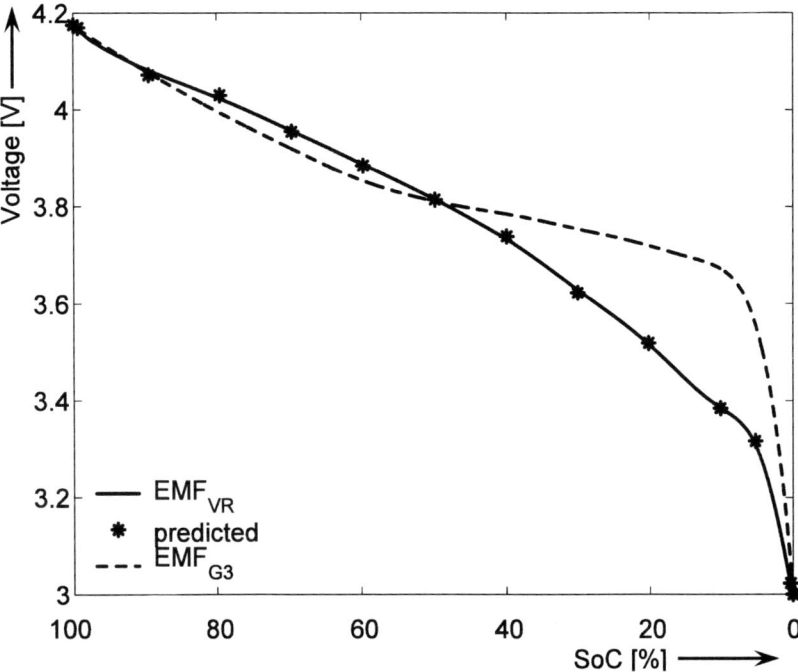

Fig. 8.7. EMF at 25°C as a function of the SoC [%].

The difference between the two EMFs amounts to 290 mV at 10% SoC. This implies an SoC inaccuracy of 24% when the US18500G3 battery parameter values are used in the SoC determination during equilibrium. This means that the adaptive system will also have to adapt the EMF_m parameters on the basis of a battery's chemistry.

The EMF dependence on temperature and the difference in charge/discharge EMFs as a function of temperature are shown in Figs. 8.8 and 8.9. The different type of battery is evidently less dependent on temperature and shows smaller differences in charge/discharge than the US18500G3 battery (see Figs. 4.7 and 4.16 for comparison). A difference between the charge/discharge EMF of 19 mV was for example measured at 19% SoC and 25°C. This implies a 1.5% inaccuracy in SoC. The EMF differences may be explained by hysteresis between the charge and discharge EMFs [4]–[6]. So, in order to ensure more accurate SoC and t_r indication, the temperature dependence and the effects of differences in charge/discharge will have to be considered in the EMF modelling.

The new SoC=f(EMF) model described by Eqs. (7.13–7.16) was used to fit the SoC=f(EMF) relationship to charge/discharge EMF curves obtained at three temperatures. Figs. 8.10 and 8.11 show that the modelled charge/discharge EMF curve used in the system shows a good fit with the discharge curve obtained with the reference battery tester at all temperatures.

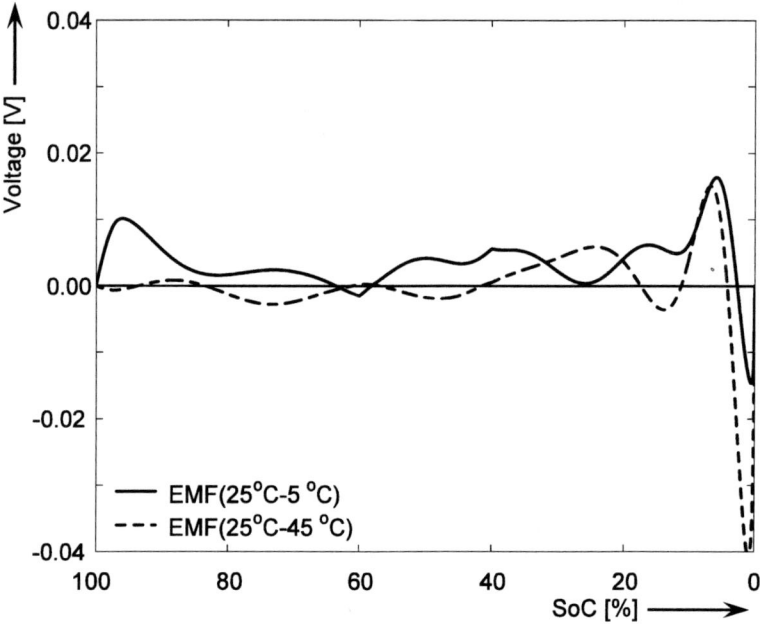

Fig. 8.8. Same as Fig. 4.7 for the new battery type.

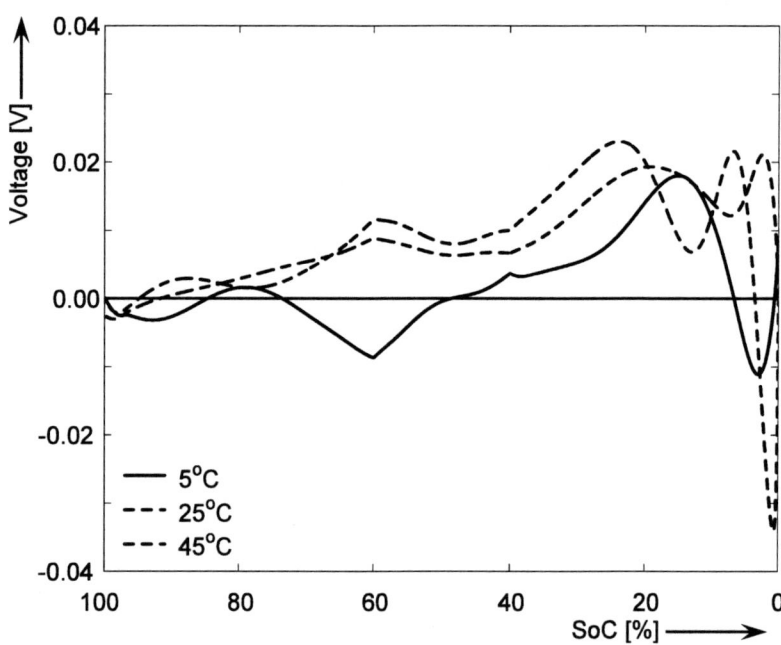

Fig. 8.9. Same as Fig. 4.16 for the new battery type.

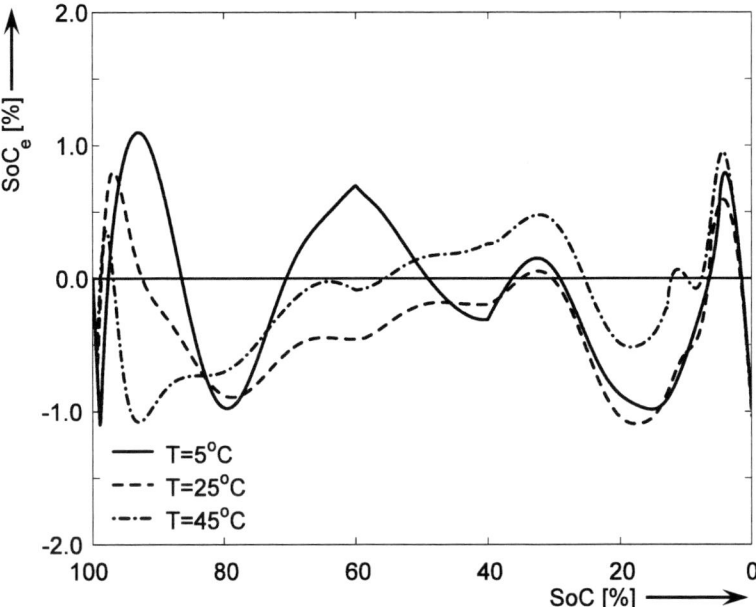

Fig. 8.10. Same as Fig. 7.8 for the new battery type.

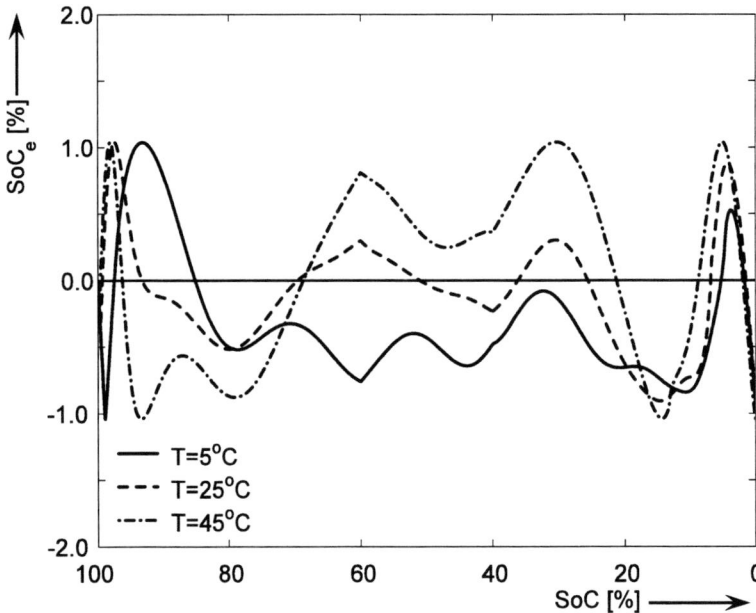

Fig. 8.11. Same as Fig. 7.9 for the new battery type.

Figs. 8.10 and 8.11 show that a maximum error in SoC of 1.1% was obtained. So the newly developed adaptive SoC=f(EMF) method enables accurate SoC calculation during equilibrium in the case of other types of batteries, too. The EMF model parameter values used in the simulations illustrated in Figs. 8.10 and

8.11 are summarised in Table 8.5 (see also Table 7.2). The value of the EMF parameter obtained for the US18500G3 battery is also given for comparison.

Table 8.5. The SoC=f(EMF) model parameter values.

Parameter	Charge value of different battery	Discharge value of different battery	Charge value of US18500G3	Discharge value of US18500G3	Unit
E_o^*	3.35	3.34	3.72	3.71	[V]
ΔE_0	$3.91\ 10^{-4}$	$3.88\ 10^{-4}$	0.00	$-3.52\ 10^{-4}$	$[V\ T^{-1}]^+$
a_{10}	$4.24\ 10^2$	$4.24\ 10^2$	2.42	2.04	[1]
Δa_{10}	$-3.24\ 10^1$	$-3.24\ 10^1$	$-4.98\ 10^{-2}$	$-4.26\ 10^{-2}$	$[T^{-1}]^+$
a_{11}	$-4.23\ 10^2$	$-4.23\ 10^2$	$2.28\ 10^{-1}$	$2.41\ 10^{-1}$	[1]
Δa_{11}	$3.22\ 10^1$	$3.22\ 10^1$	$2.28\ 10^{-3}$	$8.22\ 10^{-3}$	$[T^{-1}]^+$
a_{12}	$-1.03\ 10^{-3}$	$-1.03\ 10^{-3}$	$3.06\ 10^{-2}$	$2.05\ 10^{-2}$	[1]
Δa_{12}	$-4.6\ 10^{-7}$	$-1.08\ 10^{-6}$	$-3.83\ 10^{-4}$	$-2.46\ 10^{-4}$	$[T^{-1}]^+$
a_{20}	8.73	8.73	3.82	3.74	[1]
Δa_{20}	$-1.13\ 10^{-2}$	$-1.12\ 10^{-2}$	$2.11\ 10^{-3}$	$3.99\ 10^{-3}$	$[T^{-1}]^+$
a_{21}	$-7.81\ 10^{-1}$	$-7.81\ 10^{-1}$	$-7.59\ 10^{-1}$	$-7.59\ 10^{-1}$	[1]
a_{22}	$1.89\ 10^{-4}$	$1.91\ 10^{-4}$	$2.46\ 10^{-4}$	$6.77\ 10^{-4}$	[1]
p_{11}	$2.82\ 10^{-4}$	$1.29\ 10^{-4}$	1.29	1.35	[1]
Δp_{11}	$-2.04\ 10^{-6}$	$-2.92\ 10^{-6}$	$1.742\ 10^{-2}$	$3.58\ 10^{-3}$	$[T^{-1}]^+$
p_{12}	3.00	3.00	3.00	3.00	[1]
Δp_{12}	$6.47\ 10^{-3}$	$6.51\ 10^{-3}$	$1.86\ 10^{-2}$	$1.85\ 10^{-2}$	$[T^{-1}]^+$
p_{21}	1.04	1.04	1.06	1.06	[1]
Δp_{21}	$1.53\ 10^{-5}$	$1.52\ 10^{-5}$	$-8.49\ 10^{-5}$	$-8.01\ 10^{-5}$	$[T^{-1}]^+$
p_{22}	2.00	2.00	2.00	2.00	[1]
q_{11}	0.00	0.00	0.00	0.00	[1]
q_{12}	1.00	1.00	1.00	1.00	[1]
q_{21}	1.00	1.00	1.00	1.00	[1]
q_{22}	0.00	0.00	0.00	0.0	[1]
A	$3.41\ 10^2$	$3.41\ 10^2$	6.64	6.41	[1]
ΔA	$6.98\ 10^{-1}$	$6.81\ 10^{-1}$	$2.813\ 10^{-3}$	$3.27\ 10^{-3}$	$[T^{-1}]^+$
w_2	1.00	1.00	$9.56\ 10^{-1}$	$9.56\ 10^{-1}$	[1]
Δw_2	$-2.35\ 10^{-6}$	$-2.51\ 10^{-6}$	$-1.62\ 10^{-4}$	$-1.92\ 10^{-4}$	$[T^{-1}]^+$

[+] See Eq. (7.15).

[*] E_o^x and E_o^z have been combined. The value obtained is represented by E_o.

It follows from Table 8.5 that the charge and discharge EMF parameter values obtained for the different type of battery both differ from those obtained for the US18500G3 battery. Another conclusion relates to the parameter values calculated for the charge/discharge EMF. The charge EMF parameters obtained for the different battery are almost the same as those obtained for the discharge EMF. This leads to the conclusion that the hysteresis effect is less pronounced in the different battery. This indeed agrees with the measured data presented in Figs. 8.8 and 8.9.

SoC_l measurements and fitting

To obtain information on SoC_l the different battery was discharged from different SoC_{st} values at different constant C-rates and temperatures. After each discharge a two-hour rest period was applied. The voltage-relaxation model and the adaptive SoC=f(EMF) model were used to obtain the SoC_l. The measured SoC_l values were used as input for the SoC_{lm} described by Eq. (7.16). The result of the measured (SoC_{lm}) and fitted SoC_l (SoC_{lf}) values is presented in Fig. 8.12. The difference between the measured and fitted SoC_l values is indicated in Fig. 8.13 to show how close the measured and fitted data are to one another.

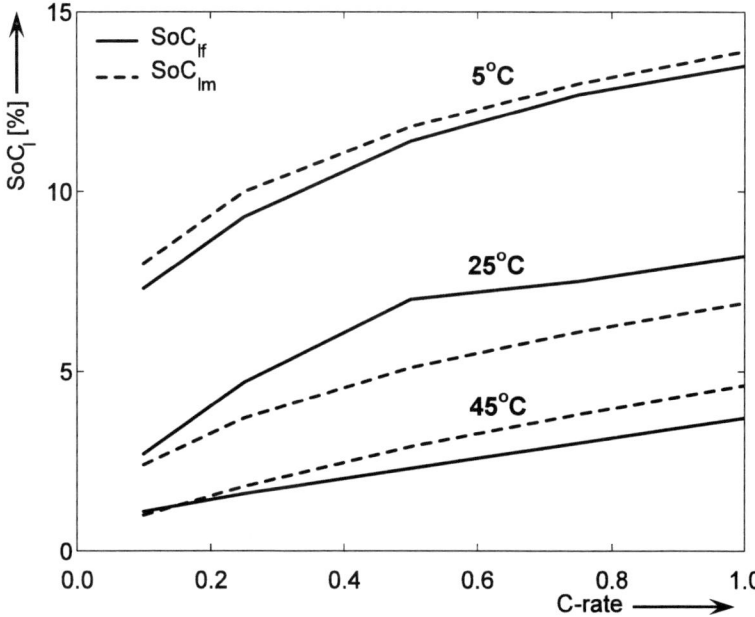

Fig. 8.12. Same as Fig. 7.10 for the new battery.

Figs. 8.12 and 8.13 show that the maximum difference between the measured and fitted SoC_l occurred at 25°C and equalled 1.8%. The influence of the SoC_{lm} inaccuracy in the t_r calculation will be discussed in the next section.

The SoC_l model parameter values used in the simulations illustrated in Figs. 8.12 and 8.13 are presented in Table 8.6 (see also Table 7.3). The values of the parameters obtained for the US18500G3 battery have been included for comparison.

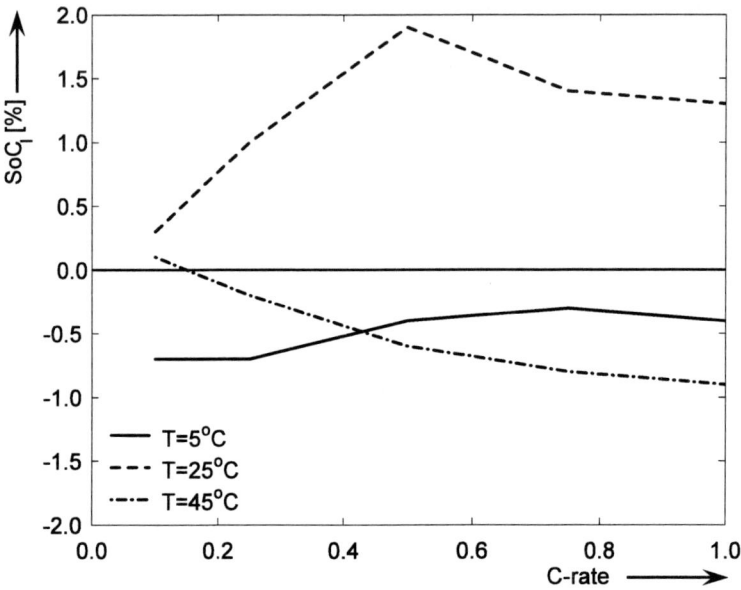

Fig. 8.13. Same as Fig. 7.11 for the new battery type.

Table 8.6. SoC-left model parameters.

Parameter	Value (different battery)	Value (US18500G3)	Unit
ς	$1.21\ 10^{-6}$	$5.90\ 10^{-6}$	–
ϑ	$2.26\ 10^{5}$	$7.25\ 10^{4}$	–
γ	$1.85\ 10^{-1}$	$3.50\ 10^{-1}$	–
δ	$1.09\ 10^{-2}$	$1.30\ 10^{-3}$	$[T^{-1}]^{+}$
α	$5.35\ 10^{-2}$	$1.18\ 10^{-1}$	–
β	$3.63\ 10^{-3}$	$3.75\ 10^{-3}$	$[T^{-1}]^{+}$

[+] See Eq. (7.16).

8.6.2 Experimental results

The different type of battery was tested in a set of tests similar to those whose results are presented in Tables 8.2 and 8.3 at constant and switching discharge C-rate currents. The results are summarised in Tables 8.7 and 8.8. Figs. 8.14 and 8.15 present t_{re} [min.] as a function of t_{rstp} [min.] to show how close the measured data presented in Tables 8.7 and 8.8 are to the actual data.

It can be concluded that the newly developed adaptive system yields a maximum error in remaining run-time prediction of –2.8%, or 4.2 minutes. The worst-case inaccuracy was obtained at 0.5 C-rate current and 45°C. In most cases the error in remaining run-time indication was however very close to the goal of 1 minute, or 1% (see Table 8.7).

Universal State-of-Charge indication for battery-powered applications 197

An uncertainty diagram similar to that presented in Fig. 8.6 holds for the different type of battery, too. In this case the ACU will adapt the system model parameters on the basis of the battery's chemistry. This means that the same adaptive models as used for the battery aging process can be used to adapt the system models to any type of battery chemistry.

Table 8.7. Same as 8.2 for the new battery type.

C-rate current	T [°C]	SoC_{st} [%]	SoC_l [%]	t_{rstp} [min]	t_{rstm} [min]	t_{re} [min]	t_{rre} [%]	SoC_{end} [%]
0.10	5	97.5	9.0	588.9	589.7	−4.5	−0.8	8.3
		67.2	7.8	395.3	395.3	0.1	0.0	7.6
		73.3	7.6	437.2	437.4	−0.7	−0.2	7.5
	25	99.9	3.2	643.5	642.9	3.8	0.6	3.7
		64.0	3.0	405.9	405.3	2.3	0.6	3.4
	45	100.0	1.3	656.8	657.3	−3.3	−0.5	0.8
		48.2	1.1	313.4	315.4	−6.3	−2.0	0.1
0.25	5	97.6	11.1	230.2	230.1	0.3	0.1	10.0
		67.0	9.6	152.8	151.7	1.6	1.1	10.0
		73.1	9.7	168.8	168.6	0.3	0.2	9.8
	25	99.9	4.6	253.7	253.1	1.5	0.6	5.2
		64.0	4.5	158.4	157.6	1.3	0.8	4.9
	45	100.0	2.3	260.1	261.5	−3.8	−1.4	0.9
		48.1	2.0	122.7	125.9	−4.0	−3.2	0.5
0.50	5	98.4	11.8	115.3	114.6	0.8	0.7	12.2
		67.0	11.5	73.9	72.9	0.7	1.0	12.0
		73.1	11.5	82.0	81.8	0.2	0.2	11.6
	25	99.9	6.2	124.7	124.0	0.9	0.7	6.9
		52.7	5.8	62.4	63.5	−0.7	−1.1	5.2
	45	100.0	3.5	128.4	130.5	−2.8	−2.1	1.5
		48.1	3.1	59.9	64.1	−2.7	−4.2	1.1
0.75	5	98.4	13.0	75.8	75.1	0.5	0.7	13.4
		67.0	12.6	48.3	47.5	0.4	0.8	13.0
		73.4	12.8	53.8	53.2	0.3	0.6	13.1
	25	99.9	7.3	82.2	81.3	0.7	0.9	8.1
		52.8	6.8	40.8	41.8	−0.4	−1.0	6.4
	45	100.0	4.5	84.7	87.3	−2.2	−2.6	2.1
		48.1	4.0	39.1	44.0	−2.0	−4.9	1.7
1.00	5	98.4	13.9	56.2	55.5	0.4	0.7	14.3
		66.8	13.2	35.7	35.7	0.0	0.0	13.4
		73.5	12.9	40.3	39.3	0.4	1.0	13.5
	25	99.9	8.2	61.0	60.3	0.4	0.7	8.9
		52.8	7.7	30.0	31.0	−0.3	−1.0	7.2
	45	100.0	5.5	62.9	65.8	−1.9	−2.9	2.6
		48.1	4.8	28.8	34.7	−1.8	−5.9	2.1

Table 8.8. Same as 8.3 for the new battery type.

Initial C-rate current	Final C-rate current	T [°C]	SoC$_{st}$ [%]	SoC$_l$ [%]	t$_{rstp}$ [min.]	t$_{re}$ [min.]	t$_{rre}$ [%]	SoC$_{end}$ [%]
0.10	0.20	5°C	67.6	7.2	401.9	3.2	0.8	9.5
0.10	0.20		73.5	7.2	441.2	2.7	0.6	9.5
0.10	0.50		98.5	7.4	606.2	1.6	0.3	12.5
0.10	1.00		98.5	7.4	606.2	0.9	0.1	14.7
0.25	0.35		67.0	9.2	153.9	1.8	1.2	10.9
0.25	0.35		73.3	9.2	170.6	1.5	0.9	10.8
0.50	0.40		67.0	10.9	74.7	1.3	1.8	11.0
0.50	0.40		73.2	11.0	82.8	1.2	1.5	11.1
0.50	1.00		98.5	11.3	116.1	0.9	0.8	14.7
0.75	0.65		67.0	12.2	48.6	0.7	1.5	12.3
0.75	0.65		73.2	12.2	54.1	0.6	1.1	12.3
1.00	0.50		98.3	13.4	56.5	1.1	2.0	11.9
1.00	0.90		67.0	13.2	35.8	0.5	1.4	13.2
1.00	0.90		73.2	13.2	39.9	0.4	1.0	13.2
0.10	0.20	25°C	63.8	3.0	404.6	2.3	0.6	4.7
0.10	0.50		99.9	3.2	643.5	1.8	0.3	7.5
0.10	1.00		99.9	3.2	643.5	0.9	0.1	9.5
0.25	0.35		52.7	4.7	127.8	−1.2	−0.9	4.3
0.50	0.1		99.9	6.2	124.7	3.0	2.5	3.6
0.50	0.40		52.7	5.8	62.4	−1.0	−1.6	4.6
0.50	1.00		99.9	6.2	124.7	0.6	0.5	9.2
0.75	0.65		52.7	6.8	40.7	−0.4	−1.0	5.9
1.00	0.50		99.9	8.2	61.0	1.1	1.8	6.9
1.00	0.90		52.8	7.7	30.0	−0.3	−1.0	6.9
0.10	0.20	45°C	48.1	1.1	312.8	−3.1	−1.0	0.8
0.25	0.35		48.1	2.0	122.7	−2.9	−2.3	1.0
0.50	0.40		48.1	3.1	59.9	−2.8	−4.5	1.4
0.75	0.65		47.9	4.1	38.9	−2.2	−5.2	1.9
1.00	0.90		48.1	4.8	28.8	−1.8	−5.8	2.4

Universal State-of-Charge indication for battery-powered applications 199

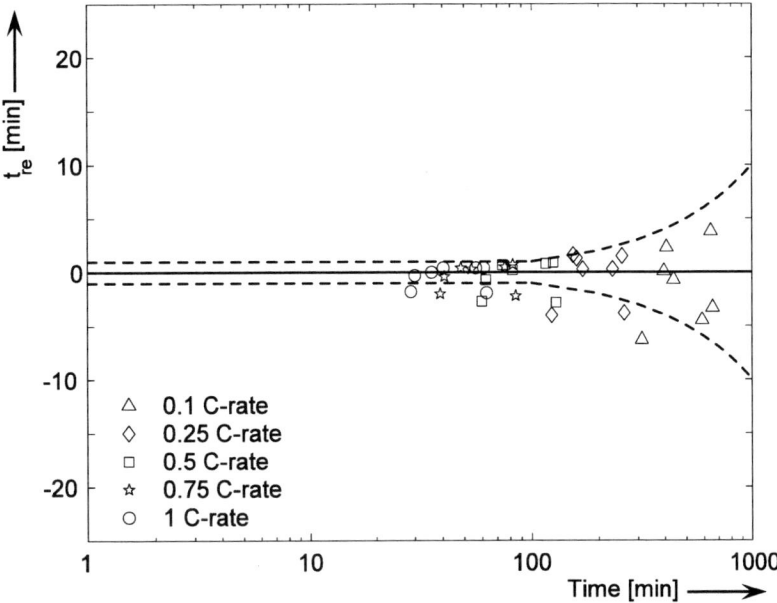

Fig. 8.14. Same as Fig. 8.4 for the new battery chemistry type.

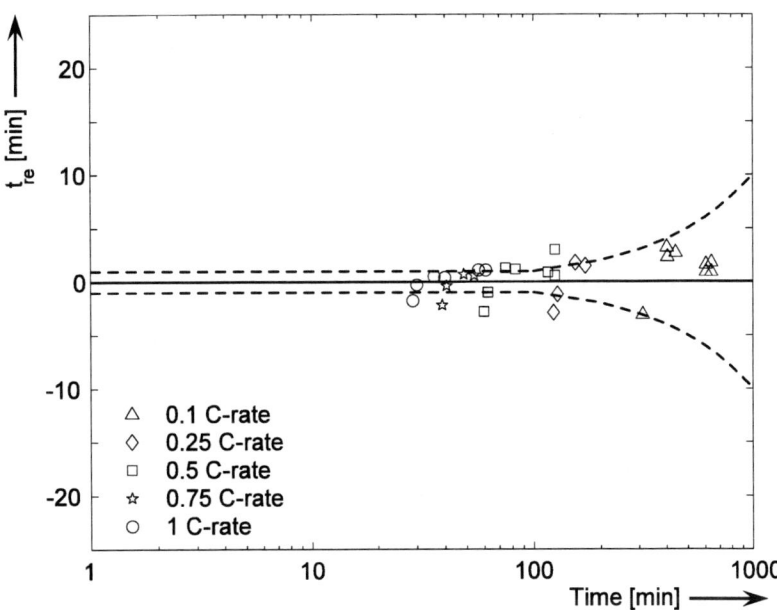

Fig. 8.15. Same as Fig. 8.5 for the new battery type.

8.7 Practical implementation aspects of the SoC algorithm

Designers will also be interested in the implementation requirements of the mathematical functions in a practical application. Close quantitative agreement with the results of laboratory simulations using the battery models and measurements in a real-time system in which the SoC system is implemented is of course important. The SoC indication system was first implemented in a real-time SoC evaluation system to test and evaluate it. The optimised SoC system's implementation had to agree with a portable device's hardware speed and memory requirements. An evaluation board on a portable device platform was therefore developed specially for this purpose.

One possible application of accurate SoC information would be for controlling charging. A battery would then always be correctly charged, implying a longer cycle life. The SoC indication system's suitability for a new ultra-fast recharging algorithm for Li-ion batteries, *i.e.* boostcharging [7], will be discussed below.

8.7.1 Hardware design of the evaluation board

The hardware required for the SoC indication algorithm's implementation in a portable device platform will now be presented. To enable flexible hardware implementation, two boards were designed for the SoC algorithm: a measurement evaluation board and a controller board (see Fig. 8.16). A connectivity block ensures the connection between these two boards.

Fig. 8.16. The measurement evaluation board (bottom right) and the controller board (top left).

The measurement evaluation board

The evaluation board contains different measurement, charging and discharging circuits. With these circuits different battery voltage, current and temperature measurement results can be compared to evaluate the SoC calculation accuracy. The evaluation board consists mainly of four blocks: *a connectivity block* used for connection to the controller board, *a current protection block* that prevents shortcircuits and high current flows, *a measurement block* to perform the voltage, current and temperature measurements and *a charging* and *discharging block* used to charge and discharge the battery. A schematic representation of the measurement circuit is shown in Fig. 8.17. There are three measurement ADC circuits: PCF50606 [8], ADS1256 [9] and ADE7759 [10]. *Rsense1* and *Rsense2* are 20 mΩ sense resistors used for the current measurement.

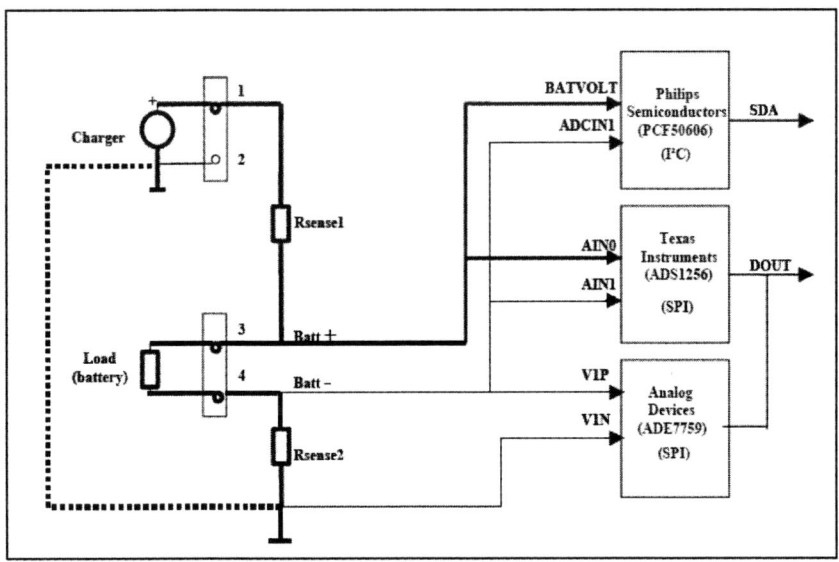

Fig. 8.17. Schematic representation of the measurement circuit connections on the evaluation board.

A charger may be connected between pins 1 and 2 and a positive current will flow through the battery connected between pins 3 and 4 (see Fig. 8.17). The PCF50606 and/or the ADS1256 may be used to measure the voltage. The ADS1256 measures the differential voltage between Batt+ and Batt-, whereas the PCF50606 measures only the potential of Batt+. A function therefore had to be implemented for the PCF50606 voltage measurement. With this device the battery voltage will be calculated as the potential difference between the *BATVOLT* and *ADCIN1* inputs (see Fig. 8.17).

The PCF50606 and/or the ADE7759 may be used for the current measurements. The current is measured as the voltage drop across a sense resistor, i.e. *Rsense1* or *Rsense2*. The value of the current is calculated by dividing the measured battery voltage by the value of the sense resistor. The PCF50606 measures only the value of the current, whereas the ADE7759 can also detect the direction of the current.

The PCF50606 is also used for temperature measurements. The battery temperature is measured using an external Negative-Temperature-Coefficient (NTC) component, placed near to the battery. The main characteristics of the ADCs chosen for the V, T and I measurements, *i.e.* PCF50606 and ADE7759, will be discussed below.

Philips Semiconductors' PCF50606

The PCF50606 is an integrated solution for power-supply generation, battery management including charging, and dedicated multi-media phone functions. The device is controlled by a host controller through an Inter-Integrated Circuit (I^2C) serial interface. On the evaluation board presented here the PCF50606 is used to measure battery voltage and temperature. Only the PCF50606 ADC module will therefore be described here. For more information on the PCF50606 the reader is referred to [8].

The ADC module consists of a 10-bit ADC with internal sample-and-hold, an input multiplexer and two high-voltage divider and subtractor circuits. The ADC is a 10-bit resistive successive-approximation converter combined with an input multiplexer and a track-and-hold circuit. The input multiplexer allows conversion of 10 different inputs. The track-and-hold circuit ensures stable input voltages at the input of the ADC during the conversion. Two inputs of the ADC multiplexer are used to measure the battery voltage applied to the BATVOLT input. The ADC module allows different settings for the full-scale input voltage. A full-scale input voltage of 3–5.4 V was chosen for the battery voltage and temperature measurements discussed here.

Analog Devices' ADE7759

The main scope of this ADC is to measure the current that flows into or out of a battery. The ADE7759 is an active power and energy measurement IC, with a serial interface and a pulse output [10]. The ADE7759 incorporates two second-order sigma-delta ADCs, a Coulomb counter, a reference circuit, a temperature sensor and all the signal processing required to perform active power and energy measurements. The Coulomb counter can be switched off if the ADE7759 is used with a current sensor.

The ADC is composed of two parts, *i.e.* a sigma-delta modulator and a digital low-pass filter. The sigma-delta modulator converts the input signal into a continuous serial stream of '1's and '0's, at a rate determined by the sampling clock. The 1-bit DAC in the feedback loop is driven by the serial data stream. If the loop gain is high enough, the average value of the DAC output (and therefore the bit stream) will approach that of the input signal level. The averaging is performed in the second part of the ADC, the digital low-pass filter. By averaging a large number of bits from the modulator, the low-pass filter can produce 20-bit data-words that are proportional to the input signal level. ADE7759 has two fully differential voltage input channels. The maximum differential input voltage for input pairs is ±0.5 V. Each analog input channel has a PGA (Programmable Gain Amplifier) with possible gain selections of 1, 2, 4, 8, and 16. In addition to the PGA, Channel 1 also has a full-scale input range selection for the ADC. By using the gain register, the maximum analog input voltage can be set to 0.5 V, 0.25 V or 0.125 V [10]. A range of +/–0.125 was chosen for the current measurements performed in the context of the research discussed here. With a sense resistor of 20 mΩ this implies a current range of +/– 6.25 A.

The ADE7759 needs to be attached to an external sampling clock. On the evaluation board presented in this chapter the ADE7759 clock is provided by the ADS1256 IC. The ADS1256 and ADE7759 circuits were to this end connected as shown in Fig. 8.18. The internal clock of the ADS1256 is attached to an 8 MHz Crystal Oscillator. The ADS1256 has a Clock Out, which generates a digital clock with the same frequency as the Crystal Oscillator. This clock can be attached to the ADE7759.

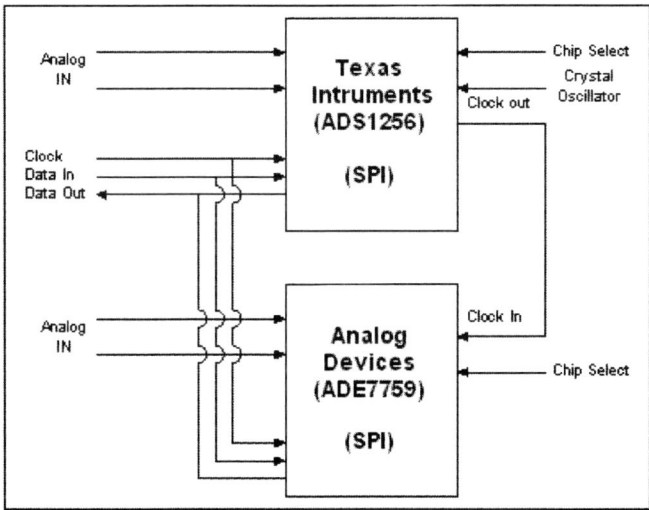

Fig. 8.18. Schematic representation of the clock connectivity.

The General Evaluation Controller Board

The controller board is used to control the evaluation board and to ensure communication with a computer. The SoC and t_r are also calculated on this board. The battery voltage, temperature and current measured with the evaluation board are the inputs for the algorithm calculations. These analog variables are digitized and fed to the controller. The Advanced RISC Machine (4ARM) [11] controls the hardware of the evaluation board through two connectors on the back of the board. These connectors are connected to the evaluation board via *the connectivity block*. The board is controlled by the LPC2292 microcontroller [12], which is suitable for control applications. It has two universal asynchronous receiver transmitter (UART) interfaces, two Serial Peripheral Interface (SPI) busses and one I^2C bus. The controller board also has a power supply and a debugging device, *i.e.* a Joint Test Action Group (JTAG) Macraigor Wiggler. A switch is used to set the board to In-System Programming (ISP) mode or In-Functional (IF) mode.

8.7.2 Software design of the evaluation board

In this section the software required for the evaluation and controller boards will be briefly presented. Several software and development tools were used to develop the software drivers for the evaluation and controller boards.

Real-Time Operating System

Due to the complex hardware design, real-time programming was needed for the software implementation. The majority of the circuits would moreover have to run at the same time, implying a need for a multitasking implementation mechanism. The most efficient way of realizing this would be by using a Real-Time Operating System (RTOS). Therefore, FreeRTOS [13] was chosen for the software implementation. FreeRTOS is a portable open source RTOS for embedded devices which provides the following features: two types of scheduling policies, *i.e.* pre-emptive (will always perform the highest available task) and cooperative (context will switch only if a task blocks), message queues, semaphores, trace visualization ability, *etc*. FreeRTOS allows tasks (parts of code that perform specific duties) to run quasi-concurrently. The tasks will appear to run all at the same time, performing specific jobs simultaneously. The scheduler will decide which task will be executed at a particular moment. For more information on FreeRTOS features, see [13].

The Integrated Development Environment (CrossWorks)

CrossWorks for ARM is a complete C-development system for ARM-7 microprocessors such as the LPC2292. It comprises an ARM C compiler, a CrossWorks C Library and a CrossStudio integrated development environment. CrossWorks is capable of flashing the software directly into the microprocessor by using JTAG, visualising input/output registers and debugging in flash. In the present application CrossStudio was used for debugging and building the project. For more information on the CrossWorks features the reader is referred to [14].

The Borland Delphi development environment

For simple communication and control a Graphical User Interface (*GUI*) was designed in Borland Delphi. In this development environment a comport-component is used to communicate with the controller board via a computer's communication port. For more information on Borland Delphi, see [15].

Software implementation

In order to offer high implementation flexibility the software structure was divided into layers. A general diagram of the software implementation is shown in Fig. 8.19.

The software application comprises four layers, *i.e.* a personal computer (PC) communication driver, an application layer and device and hardware communication driver layers. These parts each consist of one or more modules, which have their own task(s). These modules will be briefly explained below.

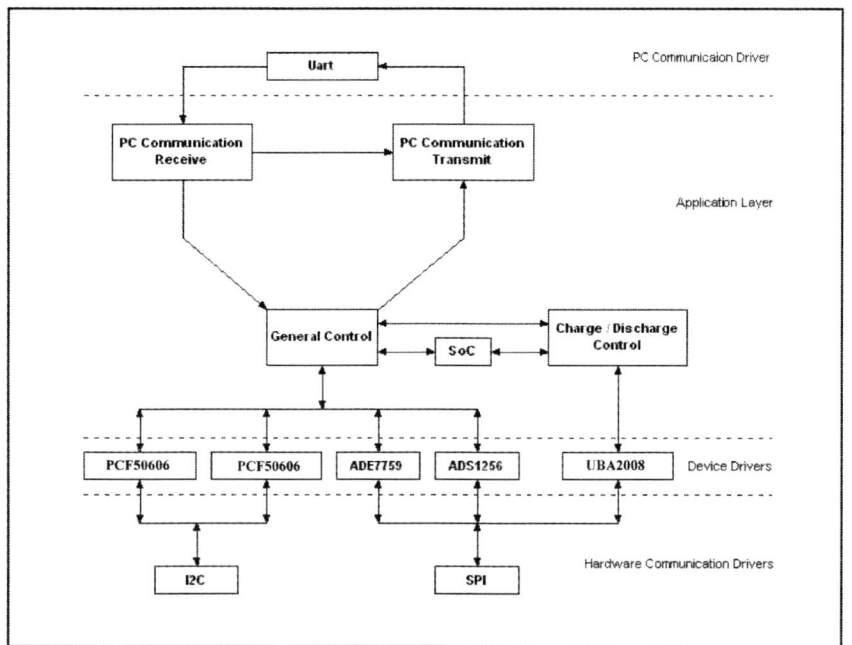

Fig. 8.19. General diagram of the software implementation.

The general control module represents the basis of the software application receiving all the notifications from the PC user input, measurement-ready ADCs, *etc*. When an event is received the general control module determines what action should be taken, *e.g.* calculate SoC by means of a particular function. The UART module enables communication with an external system. The commands given by the PC via the UART are received by the PC communication receive module. On the basis of the validity of these commands the PC communication transmit module will reply (ACKNOWLEDGE) or not reply (NACKNOWLEDGE). SoC and t_r are calculated by the SoC module by means of the algorithm developed in chapters 7 and 8. The PCF50606 and ADE7759/ADS1256 modules control the measurement circuits via the I²C and SPI module. Because PCF50606 is available on both the evaluation and the controller board, two PCF50606 modules are represented in Fig. 8.19. Finally, the UBA2008 module contains all the functions needed to set up the P89LPC932 [16] on the evaluation board and the function of the charge/discharge module is to charge or discharge the battery.

8.7.3 Measurement results

The battery voltage, temperature and current measurements obtained with the evaluation board are given as input to the SoC algorithm implemented on the controller board. The testability of the SoC algorithm models on a portable device platform will be discussed in this section. The implemented SoC algorithm combines Coulomb counting with the SoC=f(EMF) relationship, the SoC_{lm} and the Q_{max} adaptive system. But the flexible hardware and software implementation allows any other functions to be added. An example of t, V, I and T measurement and SoC and t_r calculation during discharging of a fresh Li-ion US18500G3 battery is given in Table 8.9.

Table 8.9. Battery SoC and remaining run-time calculation during discharging.

Time [min.]	V [V]	I [mA]	T [K]	SoC [%]	t_r [min.]
0.0	3.685	−277	297	12	26
0.1	3.623	−277	297	12	26
0.1	3.623	−277	297	12	26
0.1	3.621	−277	297	12	26
0.2	3.621	−277	297	12	26
0.2	3.621	−277	297	12	26
0.2	3.621	−277	297	12	26
0.3	3.621	−277	297	12	26
0.3	3.617	−277	297	12	25

During the standby mode (not illustrated in Table 8.9) the SoC algorithm calculates an SoC of 12% and a t_r of 26 minutes by means of voltage and temperature measurements and the stored SoC=f(EMF) relationship. These values correspond well to the values calculated by the SoC evaluation system discussed in chapter 7. The voltage and temperature are measured with the PCF50606. An NTC resistor was applied in the hardware and a function for calculating the temperature in the software. An NTC developed by Vishay [17] was used for this application. The resistance values of this NTC at some specific temperatures are shown in Table 8.10.

Table 8.10. NTC resistance values at different temperatures.

Temperature (°C)	Resistance (Ω)
−10	55050
5	25340
25	10000
45	4372

It should be noted that t_r during the standby state was calculated for an arbitrarily chosen 0.25 C-rate discharging current. A worst- and best-case value of t_r can be given as alternatives, with the latter representing the minimum expected load and the former the maximum expected load [3]. When discharge starts a negative current of 0.25 C-rate flows out of the battery. In this case t_r is calculated by means of cc, T measurement, SoC_{st} and SoC_{lm}. The current that flows through the battery is measured by the ADE7759 as a voltage drop across the 20 mΩ sense resistor. Every second a software function calculates the average and integral of this current. When the current, the average or the integral has been measured or calculated, the ADE7759 task will notify the General control module. A new SoC and t_r will then be calculated.

The Graphical User Interface (GUI)

A GUI was designed to control the evaluation and controller boards. The GUI enabled the voltage, current and temperature measurements to be started and the measured and calculated values to be graphically displayed on the PC screen (see Fig. 8.20). A log-file could moreover be stored in the GUI, in which the measured and calculated values could be recorded per second.

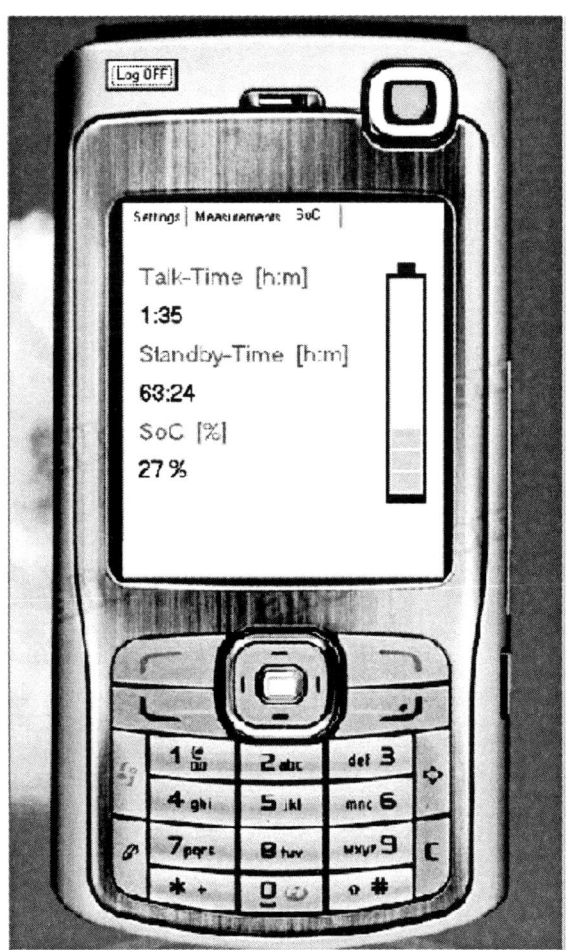

Fig. 8.20. Graphical User Interface.

The GUI screen consists of three tabs, *i.e. settings, measurements* and *SoC*. The setting tab is for the communication settings. These settings can switch the measurements on or off. The *measurements* tab displays the measured V, I, T and t values. The remaining run-time in hours and minutes [h:m], the standby time [h:m] and the SoC [%] calculations are displayed on the *SoC* tab (see Fig. 8.20). For later analysis and interpretation the measurement results and the calculations were also stored in a log-file. An example of such a log file is shown in Table 8.9.

This approach allows the developed SoC algorithm to be implemented in any battery-powered device in which the battery V, T and I are measured. The use of accurate SoC indication in a new charging concept known as boostcharging will be discussed in the next section.

8.7.4 Boostcharging

Although all small rechargeable batteries can, in principle, be frequently recharged after operation, users often regard the long time needed to recharge such high-energy batteries as rather inconvenient. Ultra-fast recharging is in many circumstances not just desirable but actually necessary, for example in the daily activities of employees working in many utility and emergency services. Rayovac proposed a new ultra-fast charging method for NiMH batteries based on hardware modifications. A similar method for Li-ion batteries would however be far more complicated.

Charging times of up to 2 hours for Li-ion batteries are not uncommon. These long charging times are mainly due to the dedicated charging algorithm that has to be used to meet the strict safety and cycle-life requirements of Li-ion batteries. It is generally accepted that deviations from conventional, standard, recharging conditions lead to side-reactions with detrimental effects on the aforementioned aspects [18], [19]. Such side-reactions are more pronounced under more extreme voltage conditions, suggesting that their frequency of occurrence is dependent on a battery's SoC. Boostcharging has been proposed as a new, ultra-fast recharging algorithm for Li-ion batteries [7], [21], [22]. Characteristic of boostcharging is that close-to-fully discharged batteries can be recharged with very high currents for a short period of time. This makes accurate SoC indication especially important in close-to-fully discharged batteries. Boostcharging Li-ion batteries and additionally charging them to full capacity in the standard way does not have any negative effects on the batteries' cycle life. Boostcharging has been shown to be very rapid. A fully discharged battery can for example be recharged to one-third of its rated capacity within 5 minutes [7].

Boostcharging characteristics will be compared with characteristics of conventional charging methods below. Close attention has been paid to possible long-term effects of boostcharging in extended cycle-life studies.

Experiments performed with the boostcharging algorithm

Boostcharging experiments were carried out using both cylindrical US18500 (Sony) and prismatic LP423048 (Philips) Li-ion batteries [7]. The results were compared with results obtained for conventionally (CCCV) charged batteries. Fresh and fully activated batteries were used for each condition to be investigated. The activation procedure was the same as that described in section 8.6. The applied boostcharge (b) regimes are defined by I_b^{max}, V_b^{max} and time (t_b). These more specific conditions will be described along with the results below. The batteries' temperatures were in all cases measured by Pt-100 sensors, which were glued directly onto the metallic casings (at mid-way positions). The experiments were performed in temperature-controlled boxes at an ambient temperature of 25°C.

Universal State-of-Charge indication for battery-powered applications

Results and Discussion

A standard CCCV charging plot for a cylindrical Li-ion battery is shown in Fig. 8.21.

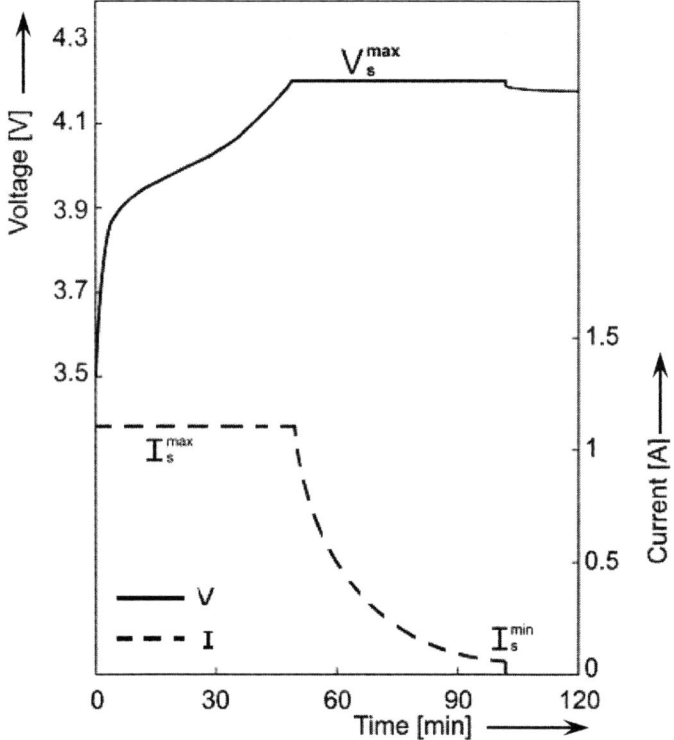

Fig. 8.21. Voltage and current characteristics of a cylindrical Li-ion cell under standard CCCV charge conditions (I_s^{max} = 1 C-rate, V_s^{max} = 4.2 V). CV charging was terminated at I_s^{min} = 55 mA (0.05 C-rate) after about 100 min.

The battery voltage clearly increased during the CC period (I_s^{max} = 1C-rate) until the V_s^{max} value of 4.2 V was reached. This caused the current to decrease in the CV region. Charging was terminated when the current cut-off value reached a current of 0.05 C-rate. Standard charging of course takes more than 100 min. In accordance with the nominal capacity, approximately 1100 mAh was delivered by the battery during the subsequent discharge period (not shown).

The simplest way of significantly accelerating the charging process is to immediately switch a battery to CV mode, i.e. without any limitations with respect to the maximum charging current. Fig. 8.22 shows such a CV-mode experiment carried out at V^{max} = 4.2 and 4.3 V. The corresponding current levels are represented in the lower part of the figure.

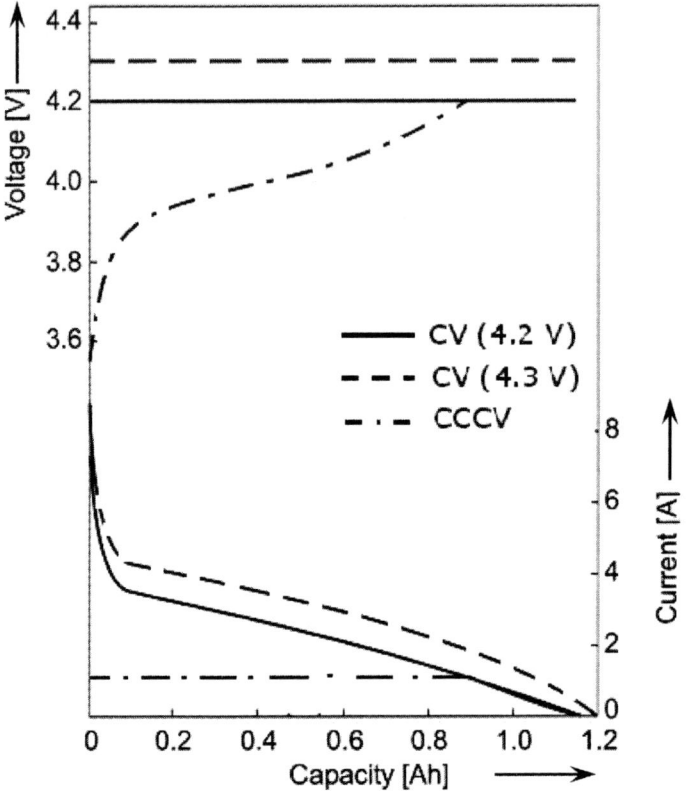

Fig. 8.22. Voltage and current characteristics of CV charging of a cylindrical Li-ion battery at V^{max} = 4.2 V and = 4.3 V, respectively. Standard CCCV charging is shown for comparison. I^{min} = 0.05 C-rate in all cases.

Initially, high charging currents of up to 8 A flowed through the batteries. This current however decreased rapidly as a result of the increasing impedance of the cells, to level off after about 2 minutes, when the capacity reached 0.04 Ah. After the initial current peak the current decreased more slowly. Standard CCCV charging results are also illustrated in the figure to allow comparison. Remarkably, the current behaviour observed in the CV-mode in both 4.2 V experiments was exactly the same. Because the current and voltage curves are generally considered to be dependent on the established concentration curves [3], this suggests that diffusion limitations do not play a significant role in this region. As expected, the currents were significantly higher at V^{max} = 4.3 V throughout the entire charging period.

Fig. 8.23 shows a comparison of the charge build-up as a function of time under the same CV and CCCV charging conditions as described above (compare with Fig. 8.22). The two distinct regions are clearly identifiable in the part representing the CCCV charging mode; during the constant-current mode the charge build-up was linear as expected. It began to deviate from this behaviour when the CV mode was entered after approximately 50 min. Such linear behaviour was of course not observed in the case of the CV-charged batteries, as constant currents were then not applied. The most striking difference between the two CV-charged batteries and the CCCV-charged battery is that observed in the charging times. 1 Ah could for example easily be obtained within 20 minutes at $V^{max} = 4.3$ V whereas this took almost 60 minutes under standard charging conditions.

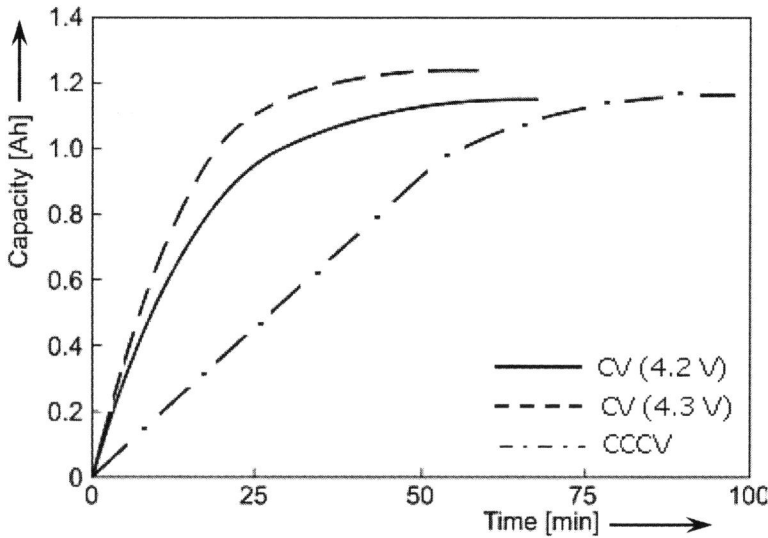

Fig. 8.23. Capacity build-up in CV charging cylindrical Li-ion batteries at 4.2 V and 4.3 V. The capacity build-up during standard CCCV charging is shown for comparison.

Fig. 8.24 shows the impact of the maximum charging current on the total charging time of an empty cell to different levels of SoC at a moderate V^{max} of 4.2 V. The graph clearly shows that hardly any time profit is gained with maximum charging currents larger than 4 A in the case of a battery of 1100 mAh. This is due to the fact that the currents initially drop very rapidly (see Fig. 8.22). Moreover, the minimum time needed to charge to a desired SoC can easily be derived from this figure. For example, it takes less than 7 minutes to charge an empty battery to 30% SoC when I^{max} is limited to 3 A at a V^{max} of 4.2 V.

The impact of the more severe CV charging conditions on the cycle life is shown in Fig. 8.25. A cycle life plot obtained under standard CCCV charging is included for comparison. Even under the moderate CCCV charging conditions the capacity loss is substantial, especially at high cycle numbers. Two regions can be clearly distinguished in the case of these cylindrical cells; in the first 300 cycles only 15 % of the original capacity is lost, whereas degradation is enhanced after this number of cycles. These two regions are attributable to two separate degradation processes, taking place at a different rate at each electrode; the slow degradation rate in the initial cycles is attributable to Li consumption in the growing SEI layer while, the high degradation rate is assumed to be attributable to decomposition of the *LiCoO₂* electrode (see Chapter 6). Allowing the maximum current to increase during CV charging at 4.2 V causes the high-rate degradation to become observable already after 160 cycles (see curve (a) of Fig. 8.25). Increasing V^{max} to 4.3 V has an even more dramatic and unacceptable effect on the cycle-life; degradation of both the graphite and the LiCoO2 electrode are significantly enhanced, as curve (b) shows. Recent reference electrode measurements performed under boostcharge conditions clearly showed that no metallic Lithium deposition takes place.

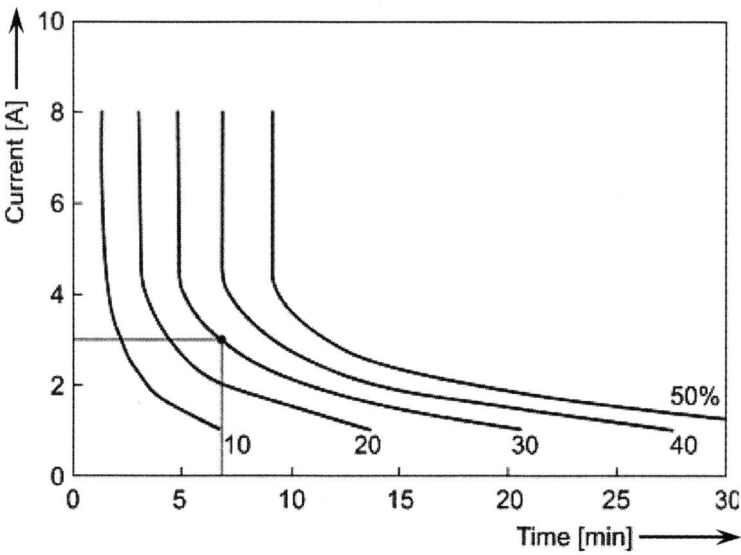

Fig. 8.24. Impact of the initial current on the total charging time to various indicated levels of SoC in cylindrical Li-ion batteries. Charging commenced at 0% SoC in all cases.

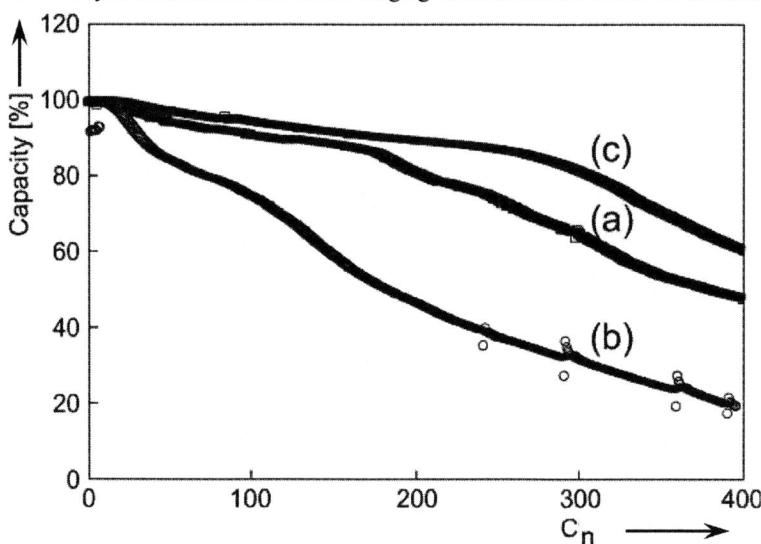

Fig. 8.25. Cycle life of cylindrical Li-ion batteries under high charging load conditions at I^{max} = 4.5 C-rate, V^{max}=4.2 V (a) and I^{max} = 4.5 C-rate, V_{max} = 4.3 V (b). Cycle life upon standard CCCV charging (I^{max}=1C, V^{max}=4.2V) is indicated by curve (c).

From the experiments described above it is clear that simply using CV charging is not feasible for ultra-fast charging of Li-ion batteries. Insight gained in recent modelling work showed that degradation is enhanced at higher SoC levels [20]. $LiCoO_2$ electrodes are assumed to be far more sensitive to decomposition into Co_3O_4 and oxygen gas at relatively low Li contents (see chapter 6). This conversely suggests that detrimental effects of this and other side-reactions can be well controlled with a new strategy in which the severe charging conditions are applied

Universal State-of-Charge indication for battery-powered applications 213

only during the initial stages of charging at low SoC values. This is why the charging algorithm referred to as "boostcharging" was introduced [7].

The basic principles of boostcharging are schematically outlined in Fig. 8.26, which represents a short boostcharge period (t_b), during which a V_b^{max} voltage is applied to a battery in the initial CV-mode.

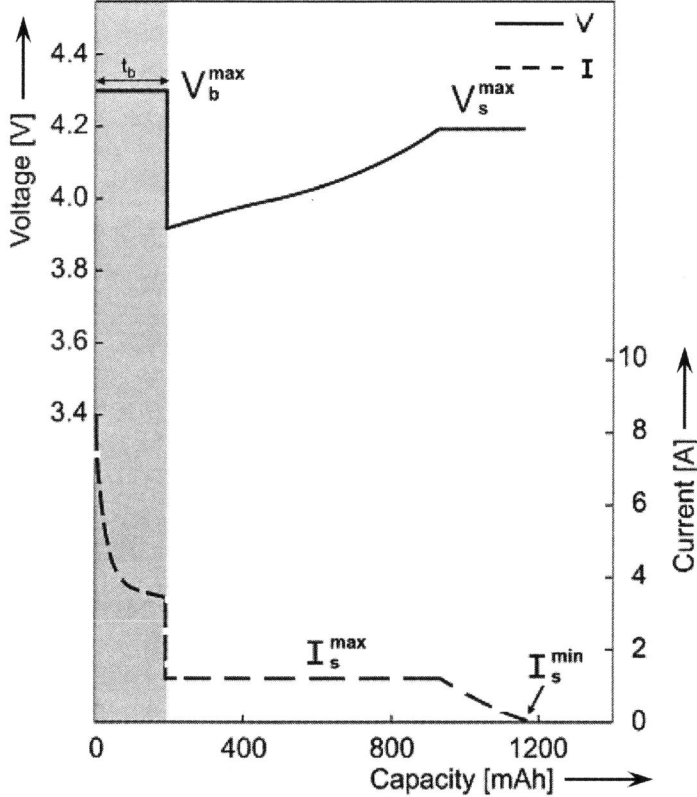

Fig. 8.26. Basic principles of boostcharging Li-ion batteries. A short boostcharge period (shaded region) is followed by standard CCCV charging. The voltage and current responses are indicated.

During the boostcharge period the currents may initially be very high. In the batteries whose results are presented in Fig. 8.22 they were up to 8 A. To ensure full charging, the boostcharging is followed by standard CCCV charging at much lower-currents. The entire boostcharge cycle can be characterised as CVCCCV charging. V_i^{max} is not necessarily the same for both CV periods as schematically indicated in Fig. 8.26. If the currents to be used in the initial boostcharging period should for some reason be unacceptably high, an alternative option could be to apply a more moderate I_b^{max} during the initial boostcharge CC-mode (not shown in Fig. 8.26). In this case, too, the boostcharging has to be followed by standard CCCV charging. The entire charging sequence can then be denoted as CCCVCCCV, or alternatively $(CCCV)^2$.

Typical boostcharge results obtained with cylindrical cells are shown in Fig. 8.27. During the relatively short boostcharge period (t_b = 5 min.) a V_b^{max} of either 4.2 V or 4.3 V was applied to the batteries. After 5 minutes the charging regime was switched to standard CCCV-mode. Conventional CCCV charging results are shown for comparison. The resulting currents are represented by the corresponding curves. The higher V_i^{max} values of course led to higher currents. Although the boostcharge period was relatively short, the capacity built up in this time was significant. This becomes apparent when the results shown in Fig. 8.27 are plotted *versus* the amount of charged capacity. This can be seen in Fig. 8.28. The lower part of Fig. 8.28 shows that the temperature rise is current-dependent, but is also fairly moderate under all boostcharge conditions. Even charging at extremely high currents of up to 5 C-rate did not cause the temperature to change by more than 10°C.

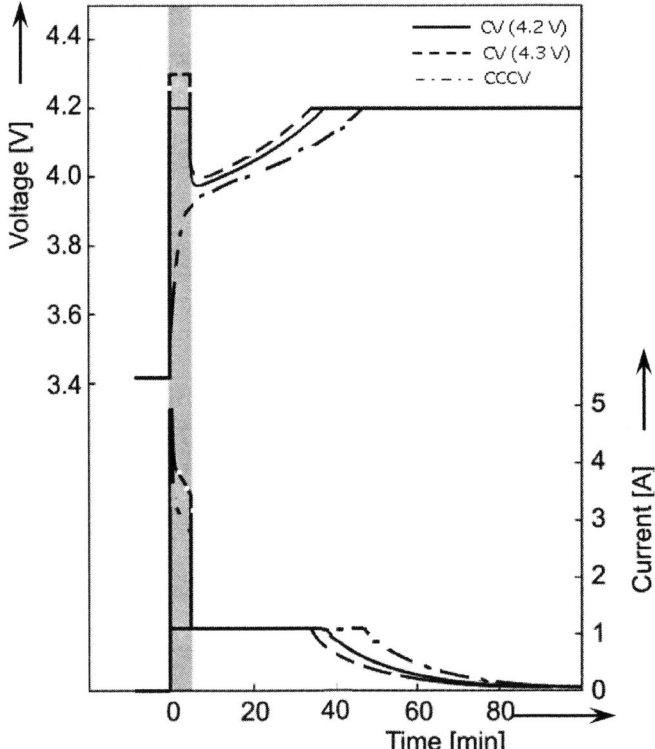

Fig. 8.27. Voltage and current transients for cylindrical Li-ion batteries under various boostcharging and (subsequent) standard charging conditions. The boostcharging was performed at 0<t_b<5 min., I_b^{max} =4.5 C-rate, V_b^{max} =4.2 V followed by standard charging; 0<t_b<5 min., I_b^{max} =4.5 C-rate, V_b^{max} =4.3 V followed by standard charging. Standard charging conditions for t>0 min. and t>5 min. are I_s^{max} =1 C-rate, V_s^{max} =4.2 V.

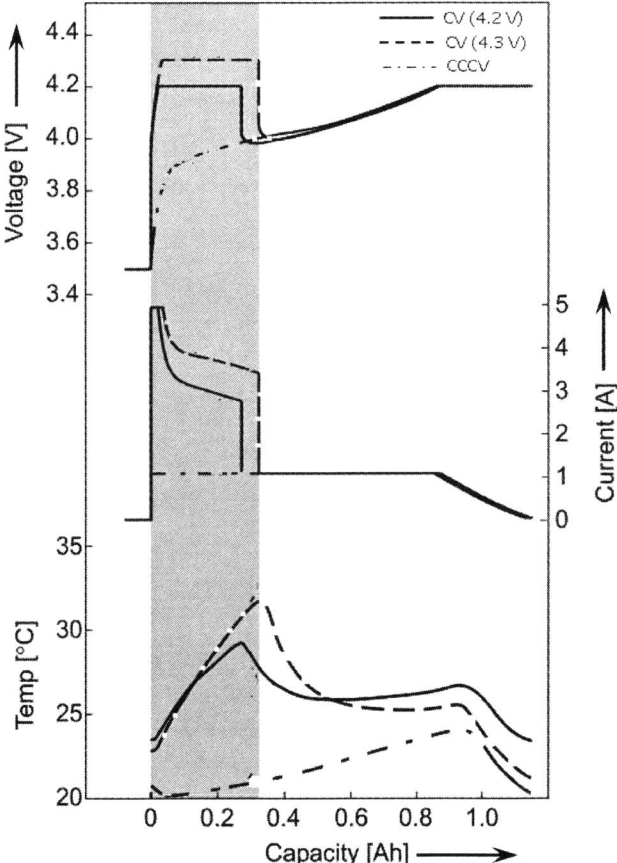

Fig. 8.28. Same results as shown in Fig. 8.27, only now presented as a function of charged capacity. Also included is the temperature development under the various boostcharging and standard conditions.

Fig. 8.29 shows that recharging can be very quick under boostcharging conditions. At 0% SoC, about 60 and 33% of the nominal capacity of a completely discharged battery can be recharged within 10 and 5 minutes, respectively, at a V_b^{max} of 4.3 V. These values are somewhat lower at lower V_b^{max} values and are very low under standard charging conditions, as is to be expected.

The impact of a battery's initial SoC condition on the boostcharge capacity is represented in Fig. 8.30 for three different boostcharging times. The boostcharge conditions are $V_b^{max} = 4.3$ V at $I_b^{max} = 5$ A. These results clearly show that boostcharging is most effective at lower initial SoC values.

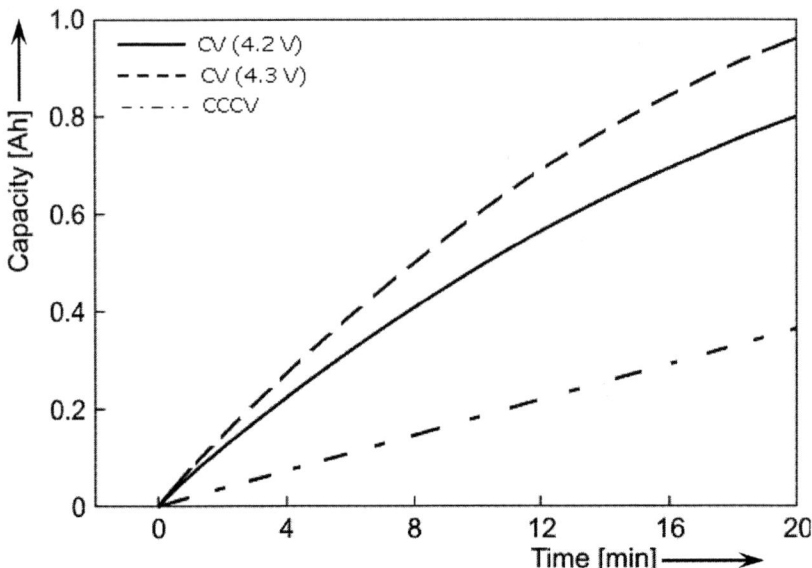

Fig. 8.29. Capacity build-up during boostcharging of cylindrical Li-ion batteries at 4.2 V and 4.3 V. The capacity build-up during standard CCCV charging is shown for comparison. The same boostcharging conditions were used as indicated in Fig. 8.27, except that the boostcharging time was in this case not restricted to 5 minutes.

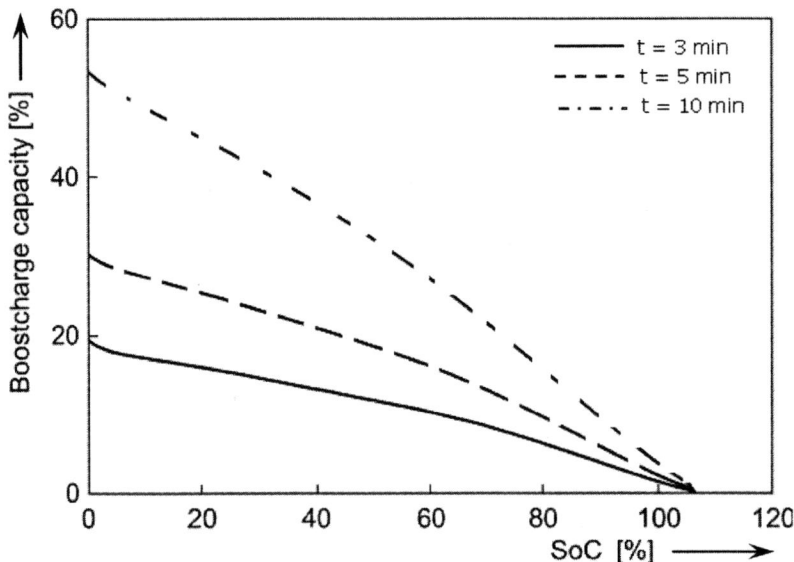

Fig. 8.30. Impact of the boostcharging time and initial SoC at which boostcharging is initiated on the boostcharge capacity: The boostcharging times were 3, 5 and 10 minutes at I_b^{max} =4.5 C-rate, V_b^{max} =4.3 V.

Fig. 8.31 shows a comparison of the cycle-life behaviour after standard charging (solid line) and boostcharging (dashed line). In the latter case the 5 minutes' boostcharging period was again followed by standard charging up to 100% SoC. It is clear that boostcharging with additional standard CCCV charging does not have any negative impact on the cycle life in comparison with standard charging, implying that degradation begins at higher SoC levels.

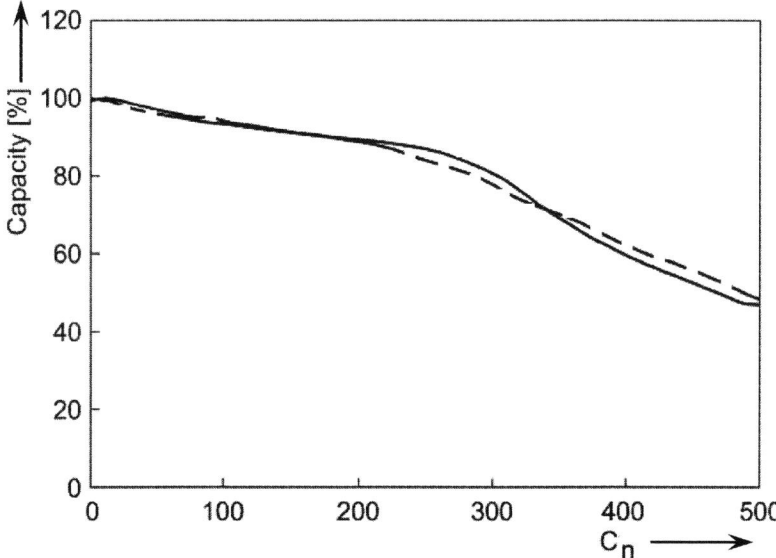

Fig. 8.31. Comparison of the cycle-life performance of cylindrical Li-ion batteries upon standard charging (solid line) and boostcharging (dashed line). The standard charging conditions were I_s^{max} =1 C-rate, V_s^{max} =4.2 V; the boostcharging conditions were 0<t_b<5 min., I_b^{max} =4.5 C-rate, V_b^{max} =4.3 V followed by standard charging.

Similar boostcharging experiments were performed with prismatic cells. Some cycle-life results obtained for the prismatic cells are shown in Fig. 8.32, in which the dashed curves correspond to V_b^{max} = 4.3 V and I_b^{max} was limited to 2.4 A (4 C-rate). The results obtained under even more extreme boostcharging conditions (V_b^{max} was raised to 4.4 V while I_b^{max} remained at the same high value) are also represented. The boostcharging consequently comprised only CC charging and the entire charging algorithm can be characterised as CCCCCV. The charge capacity under the latter conditions was somewhat higher, as was to be expected. It should be noted that battery packs are generally equipped with safety electronics, generating a significantly higher internal resistance, which makes it possible to safely recharge the battery at these "compensated" high voltage levels.

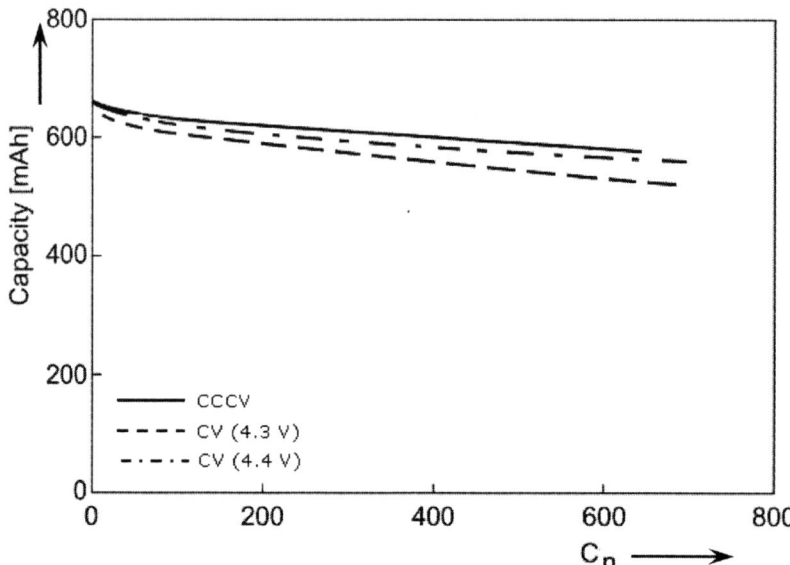

Fig. 8.32. Comparison of the cycle-life performance of prismatic Li-ion batteries upon standard charging and boostcharging. The standard charging conditions were I_s^{max} =1 C-rate, V_s^{max} =4.2 V; the boostcharging conditions were 0<t_b<5 min., I_b^{max} =4 C-rate, V_b^{max} =4.3 V and 4.4 V followed by standard charging.

The results show that boostcharging does not have a significant impact on cycle life. The observed differences are probably due to statistical variations, as the more severe boostcharging conditions led to somewhat better performance than the more moderate conditions. Also noteworthy is that no second degradation effect was observed up to cycle number 700, whereas this was the case with the cylindrical cells (compare Figs. 8.31 and 8.32).

It should be noted that in the experiments discussed in this section the boostcharging algorithm was applied mainly at low SoC levels. Applying the booscharging algorithm under the same conditions at higher SoC levels may lead to accelerated degradation of Li-ion batteries. This makes accurate SoC determination at low SoC levels very important for optimum implementation of the booscharging algorithm.

8.8 Conclusions

Adaptive systems for different types of Li-ion batteries have been tested under an extended range of conditions. The adaptive systems are suitable for a State-of-Charge algorithm that calculates the SoC in percentage as well as the remaining run-time for any battery-powered application.

Two adaptive systems have been compared. Of the two, the SoC=f(EMF) and SoC_{lm} adaptive system is recommended for SoC and t_r determination when a battery ages. An adaptive real-time SoC evaluation system has consequently been developed and tested. Significant improvements in remaining run-time prediction accuracy have been obtained with this new system. In most cases the maximum error in remaining run-time prediction was less than one minute, or 1% (see Figs. 8.4, 8.5, 8.14 and 8.15). It can be concluded that the adaptive SoC system represents the best solution for obtaining accurate SoC and t_r values.

The innovative SoC evaluation system has also been used to test a different type of battery. In those tests, new parameter values were learned by means of the previously developed adaptive systems. The results show that the SoC evaluation system is capable of adapting to batteries with different chemistries and of offering accurate, universal SoC indication.

A possible implementation of the SoC evaluation system on a mobile phone platform has also been presented in this chapter. The measured V, T, and I and calculated SoC and t_r variables obtained with this system corresponded well to the values calculated with the SoC evaluation system discussed in chapter 7, implying that the SoC system's memory and speed requirements correspond to those available in a portable application.

One possible application of accurate SoC information would be for controlling charging. In this chapter a special charging algorithm, *i.e.* boostcharging, has been presented. Boostcharging is used to recharge close-to-fully discharged batteries with very high currents for a short period of time without any detrimental effects. So accurate SoC calculation under these conditions is very important: a correctly charged battery will have a longer cycle life.

8.9 References

[1] H.J. Bergveld, V. Pop, P.H.L. Notten, Method of estimating the state-of-charge and of the use time left of a rechargeable battery, and apparatus for executing such a method, Patent WO2005085889, filed 23 February (2005)
[2] H.J. Bergveld, V. Pop and P.H.L. Notten, Apparatus and method for determination of the state-of-charge of a battery when the battery is not in equilibrium, Patent PH006795EP1, filed October 30 (2006)
[3] H.J. Bergveld, W.S. Kruijt, P.H.L. Notten, Battery Management Systems, Design by Modelling, **1**, Kluwer Academic Publishers, Boston (2002)
[4] H. Wonchull, H. Mitsuhiro, K. Tetsuichi, Hysteresis on the electrochemical lithium insertion and extraction of hexagonal tungsten trioxide. Influence of residual ammonium, Solid State Ionics Journal, **128**, 25–32 (2000)
[5] T. Zheng, W.R. McKinnon, J.R. Dahn, Hysteresis During Lithium Insertion in Hydrogen-Containing Carbons, J. Power Sources, **143**, 2137–2145 (1996)
[6] T. Zheng, J.R. Dahn, J. Power Sources, Hysteresis Observed in Quasi-open Circuit Voltage Measurements of Lithium Insertion in Hydrogen-containing Carbons, **68**, 201–203 (1997)
[7] P.H.L. Notten, J.R.G.C. van Beek, J.H.G. Op het Veld, Boost-charging Li-ion batteries: a chalenging new charging algorithm, J. of Power Sources, **145**, 89–94 (2005)
[8] Philips Semiconductors, PCF50606 Controller for Power Supply and Battery Management, Datasheet (2002)
[9] Texas Instruments, ADS1256 – Very Low-Noise, 24-Bits Analog-to-Digital (A/D) Converter, Datasheet (2004)
[10] Analog Devices, ADE7759 – Active Energy Metering IC with di/dt Sensor Interface, Datasheet, (2002)

[11] Emdes, 50-0080-1, 4ARM Emdes Layout, (2004)
[12] Philips Semiconductors, LPC2119/2129/2194/2292/2294, User Manual, (2004)
[13] Real-time Operating System, FreeRTOS, www.freertos.org
[14] Rowley associates LTD, www.rowley.co.uk/documentation/arm/index.htm, Rowley CrossWorks
[15] Delphi development environment, Borland Delphi, www.borland.com/delphi/
[16] Philips Semiconductors, P89LPC932, Datasheet (2004)
[17] Vishay BCcomponents, NTC Thermistors, Accuracy Line 2381 640 5103, Datasheet (2005)
[18] P. Arora, R.E. White and M. Doyle, J.Electrochem. Soc., **145**, 3647 (1998)
[19] M. Broussely, S. Herreyre, P. Biensan, P. Kasztejna, K. Nechev, R.J. Staniewicz, J. Power Sources, **97**, 13 (2001)
[20] D. Danilov and P.H.L. Notten, 12th International Meeting on Lithium Batteries, Nara, Japan (2004)
[21] J.R.G.C. van Beek, P.H.L. Notten, Charger for rechargeable batteries, Patent PHNL020542 (2005)
[22] J.R.G.C. van Beek, P.H.L. Notten, Charger for rechargeable batteries, Patent PHNL020541 (2005)

Chapter 9
General conclusions

State-of-Charge (SoC) and remaining run-time (t_r) indication involves battery measurements and modelling. The main focus in this book has been on designing a universal SoC system that accurately calculates the SoC in percentage and a t_r in minutes for an Li-based battery-powered device.

As stated in chapter 1, the dream of the last 70 years of research in the field of SoC has been to design a universal SoC system that adapts on-line to any type of battery without user intervention. Until this research was started, no one had succeeded in coming up with an SoC system that is accurate enough under all realistic user conditions. This book presents innovative, accurate solutions that enable on-line adaptation to different types of battery chemistry [1], [2].

Good measurement methods combined with an efficient testing scheme are the basis of developing any SoC system. Measurements are necessary to obtain accurate information on a battery's behaviour under an extended range of operating conditions and to allow the development of accurate battery models. A proper understanding of battery behaviour is necessary to improve SoC and t_r indication accuracy. Correct information on a battery's Electro-Motive Force (EMF) dependence on battery aging was for example obtained only by applying accurate maximum capacity and EMF measurement methods. Those measurement methods and newly developed adaptive systems greatly improved, SoC indication accuracy under equilibrium conditions when a battery ages.

Also important with respect to understanding battery behaviour is overpotential symmetry, a phenomenon discovered during the analysis of overpotential measurement results. The properties of this phenomenon have been used to arrive at accurate SoC determination with due allowance for aging effects.

Once a proper understanding of battery behaviour had been obtained, a physical or mathematically equivalent description in the form of a model could be built for different types of battery. The models presented in this book are consequently based on physical and electrochemical theory and the various processes can be easily recognised in them. They include a variety of parameters whose values depend on the determination method and experimental conditions. These models enable accurate simulation of battery behaviour under a wide variety of operating conditions.

The various modelling and simulation examples given in this book show that an accurate battery model is essential for proper SoC indication. A model has been developed that describes a battery's EMF and overpotential behaviour, neither of which can be measured directly. The EMF and overpotential curves were obtained in battery measurements and were implemented in an SoC evaluation system using approximation by means of mathematical functions. Accurate modelling was likewise indispensable in arriving at a new voltage-prediction model that speeds up SoC indication on the basis of the EMF during the relaxation process. This model significantly improved the calibration and adaptation possibilities of the SoC system.

A performance analysis of the SoC system was carried out to test the accuracy of the SoC algorithm. This was done by implementing the developed battery models in an SoC evaluation system. The simulated battery behaviour was checked by comparing it with measurements performed with a real-time SoC evaluation system. Close quantitative agreement between the simulation results obtained using the battery models and measurements using a real-time system based on the SoC algorithm was of course essential.

The next step was to identify the sources of error in the SoC evaluation system to arrive at more accurate SoC and remaining run-time calculation. Using uncertainty analysis for the purpose of improving SoC indication proved to be an elegant and effective approach. This led to new SoC=f(EMF) and State-of-Charge-left (SoC_l) models. In these models the SoC during equilibrium and t_r during discharge are calculated by means of one direct function. Voltage measurements and EMF calculations under current flowing conditions are not needed for the t_r calculation. This is a major advantage over previously described systems, showing that a proper understanding of sources of error allows significant SoC indication improvements.

Battery aging is a complex process that involves many battery parameters, the most important of course being storage capacity. Adaptive systems enabling accurate SoC determination when a battery ages have been developed. Adaptive models for Q_{max} and overpotential measurement are presented in this book. These models combine the advantages of the charging process with phenomena discovered during the analyses of the measurements. Measurement analyses were also used in order to develop an SoC=f(EMF) and SoC_{lm} adaptive system.

The adaptive systems were then implemented in a real-time SoC evaluation system. They were used to test fresh and aged batteries under an extended range of conditions. In the great majority of the tests the error in the predicted remaining run-time was less than one minute, or 1%, which satisfies the goal of the research described in this book.

The system must of course also yield accurate results for other types of batteries. It was therefore extended and tested on a different type of battery. The previously developed adaptive systems were used to obtain new parameter values. The results show that the SoC evaluation system is indeed capable of adapting to different battery chemistries and providing accurate, universal SoC indication.

For a proper evaluation, the system developed during this study was compared with a competitive system, Texas Instruments' book-keeping system bq26500 SoC IC. The two systems were compared by subjecting them to an identical series of tests. The comparison showed that the newly developed SoC evaluation system performs much better than the bq26500 SoC system under all conditions.

Designers will also be interested in the implementation requirements of the mathematical functions in a practical application. The optimised SoC system must agree with a portable device's hardware speed and memory requirements. The SoC models were therefore implemented in a demonstration board incorporating a hardware platform of a mobile phone. The calculated SoC and t_r values were found to agree with the values obtained in simulations, implying that the SoC system requirements correspond to those available in portable applications.

One possible application of accurate SoC information is for controlling charging. This was investigated using a newly developed, ultra-fast recharging algorithm, *i.e.* boostcharging. Characteristic of boostcharging is that close-to-fully discharged batteries can be recharged with very high currents for a short period of time without any detrimental effects. Accurate SoC calculation is very important under such conditions, for determining the conditions under which boostcharging is allowed.

References

[1] H.J. Bergveld, V. Pop, P.H.L. Notten, Method of estimating the state-of-charge and of the use time left of a rechargeable battery, and apparatus for executing such a method, Patent WO2005085889, filed 23 February (2005)
[2] H.J. Bergveld, V. Pop and P.H.L. Notten, Apparatus and method for determination of the state-of-charge of a battery when the battery is not in equilibrium, Patent PH006795EP1, filed October 30 (2006)

Philips Research Book Series

1. H.J. Bergveld, W.S. Kruijt and P.H.L. Notten: *Battery Management Systems*. 2002
   ISBN 1-4020-0832-5
2. W. Verhaegh, E. Aarts and J. Korst (eds.): *Algorithms in Ambient Intelligence*. 2004
   ISBN 1-4020-1757-X
3. P. van der Stok (ed.): *Dynamic and Robust streaming in and between Connected Consumer-Electronic Devices*. 2005   ISBN 1-4020-3453-9
4. E. Meinders, A.V. Mijritskii, L. van Pieterson and M. Wuttig: *Phase-Change Optical Recording Media*. 2006   ISBN 1-4020-4216-7
5. S. Mukherjee, E. Aarts, R. Roovers, F. Widdershoven and M. Ouwerkerk (eds.): *AmIware*. Hardware Technology Drivers of Ambient Intelligence. 2006
   ISBN 1-4020-4197-7
6. G. Spekowius and T. Wendler (eds.): *Advances in Healthcare Technology*. Shaping the Future of Medical Care. 2006   ISBN 1-4020-4383-X
7. W.F.J. Verhaegh, E. Aarts and J. Korst (eds.): *Intelligent Algorithms in Ambient and Biomedical Computing*. 2006   ISBN 1-4020-4953-8
8. J.H.D.M. Westerink, M. Ouwerkerk, Th. Overbeek, F. Pasveer, and B. de Ruyter (Eds.): *Probing Experience: from Academic Research to Commercial Propositions*. 2008
   ISBN 978-1-4020-6592-7
9. V. Pop, H.J. Bergveld, D. Danilov, P.P.L. Regtien and P.H.L. Notten: *Battery Management Systems*. Accurate State-of-Charge Indication for Battery-Powered Applications. 2008   ISBN 978-1-4020-6944-4